Urban Sustainability

The Urban Sustainability Book Series is a valuable resource for sustainability and urban-related education and research. It offers an interdisciplinary platform covering all four areas of practice, policy, education, research, and their nexus. The publications in this series are related to critical areas of sustainability, urban studies, planning, and urban geography.

This book series aims to put together cutting-edge research findings linked to the overarching field of urban sustainability. The scope and nature of the topic are broad and interdisciplinary and bring together various associated disciplines from sustainable development, environmental sciences, urbanism, etc. With many advanced research findings in the field, there is a need to put together various discussions and contributions on specific sustainability fields, covering a good range of topics on sustainable development, sustainable urbanism, and urban sustainability. Despite the broad range of issues, we note the importance of practical and policy-oriented directions, extending the literature and directions and pathways towards achieving urban sustainability.

The series will appeal to urbanists, geographers, planners, engineers, architects, governmental authorities, policymakers, researchers of all levels, and to all of those interested in a wide-ranging overview of urban sustainability and its associated fields. The series includes monographs and edited volumes, covering a range of topics under the urban sustainability topic, which can also be used for teaching materials.

Saeid Pourroostaei Ardakani · Ali Cheshmehzangi
Editors

Digital Twin Computing for Urban Intelligence

Springer

Editors
Saeid Pourroostaei Ardakani
School of Engineering and Physical Sciences
University of Lincoln
Lincoln, Lincolnshire, UK

Ali Cheshmehzangi
School of Architecture, Design and Planning
University of Queensland
Brisbane, SA, Australia

ISSN 2731-6483 ISSN 2731-6491 (electronic)
Urban Sustainability
ISBN 978-981-97-8482-0 ISBN 978-981-97-8483-7 (eBook)
https://doi.org/10.1007/978-981-97-8483-7

This Springer imprint is published by the registered company Springer Nature Singapore Pte Ltd.
The registered company address is: 152 Beach Road, #21-01/04 Gateway East, Singapore 189721, Singapore

For country, its partisans and patriots, and where we hope there is still hope.

Preface

We need to create a new urban intelligence, using digital platforms to organise local debate and give residents the ability to seize the technological future of their city.

—Bernard Stiegler, Philosopher, President of Ars Industrialis, and Director of the French Institute for Research and Innovation

This book establishes the foundational framework for an examination of Digital Twin Computing (DTC) within the realm of urban intelligence. Stretching from fundamental knowledge to technologies and application, we aim to explore beyond just methods and technologies that are already exploited in our industry and academia. In light of the ongoing surge in urbanisation, innovative methodologies are necessary for effective urban planning, infrastructure administration, and sustainable progress. DTC emerges as a promising paradigm, offering a comprehensive and data-centric approach to addressing these challenges. Hence, the importance of urban intelligence and how it could be achieved beyond the mere facades and slogans of smart city development.

Fundamentally, digital twin computing entails the development of dynamic digital duplicates, known as "twins" of physical urban systems. These real-time representations are continuously enriched with data sourced from sensors, the Internet of Things (IoT), and other channels. Through the amalgamation of diverse data streams and sophisticated analytics, this approach facilitates a thorough comprehension, predictive modelling, and optimisation of intricate urban landscapes.

The integration of DTC into urban contexts results in the formulation of Urban Digital Twins (UDTs). These encompass a multitude of facets within a city, including infrastructure, transportation, environmental factors, and social dynamics. And undoubtedly, there is an immediate need for institutional rearrangements, innovation, and adjustments so that new paradigms and policies could become what we essentially need rather than experimenting drastic changes to society. We can argue that urban digital twins serve as robust decision-support mechanisms for urban stakeholders, furnishing insights into urban dynamics, trends, and potential interventions.

A pivotal aspect of DTC for Urban Intelligence is its reliance on data-driven decision-making. UDTs harness the vast array of urban data, spanning from traffic patterns and energy consumption to air quality and social media engagements. This empowers stakeholders with informed decision-making abilities and facilitates optimised resource allocation, with applications ranging from enhancing public transportation systems to mitigating natural disasters and improving public safety protocols.

Nevertheless, realising the complete potential of DTC for Urban Intelligence necessitates the resolution of technical impediments such as data integration, interoperability, and scalability to ensure seamless functionality across diverse urban systems. Additionally, ethical considerations pertaining to data privacy, security, and mitigating algorithmic biases warrant careful deliberation to mitigate potential risks and ensure fair outcomes for all urban denizens.

Despite these challenges, the transformative capacity of DTC for Urban Intelligence is indisputable. Through fostering collaboration among urban planners, engineers, data scientists, and policymakers, we can harness the capabilities of digital twins to cultivate smarter, more resilient, and sustainable cities of the future. This narrative invites readers to delve into this transformative domain and envisage the potential advancements in urban living in the forthcoming years.

Lincoln, UK — Saeid Pourroostaei Ardakani
Brisbane, Australia — Ali Cheshmehzangi

Acknowledgements

As always, we do not want to exclude anyone's contribution to this exciting project. We would like to express our appreciation to all authors, co-authors, chapter contributors, and collaborators who have contributed their expertise, passion, and creativity to this endeavour. The same applies to our earlier work and now, valuing the contributors' dedication and vision that created this book. Such collaborative opportunity creates a new dialogue, a new networking opportunity, as well as a valuable research exchange platform for many scholars worldwide. Despite our concerns and the never-ending disruptions, we continue to create a flow that matters to the majority of us, bearing in mind there is always room for innovation and improvement, particularly when it comes to intelligence.

About This Book

Digital Twin Computing for Urban Intelligence focuses on new and ongoing discourses in interdisciplinary research and practice in urban system and smart city development pathways. It approaches digital twin fundamentals and principals including theoretical foundations, conceptualisations, strategies, and services/patterns to define and adapt digital twin solutions for urban applications—mainly sustainability.

This book highlights promising case studies and outlines digital twin design models and system architecture by examining key digital twin deployment practices such as data analysis, decision-making, and service automation in line with intelligent urban planning. It also emphasises on DT technologies such as cloud computing, AI, IoTs, and smart virtualisation and outlines the key benefits of the DT solutions in urban applications—mainly control and planning.

This book is intended for a wide range of audiences, including interested layperson audiences, undergraduate and graduate students in university, and researchers. The key benefits of this book are:

1. To introduce the theoretical principles and fundamentals of DT computing for urban intelligence.
2. To present the state-of-the-art DT technologies in urban planning and control.
3. To describe the practical deployment process of DT computing solutions for urban sustainability and intelligence.

Contents

About the Editors

Saeid Pourroostaei Ardakani currently works as a Senior Lecturer in Computer Science at the University of Lincoln, UK. He is also a member of MLearn research group at the University of Lincoln, and has formerly worked at the University of Nottingham (UNNC) and Allameh Tabatabai University (ATU) as an Assistant Professor in Computer Science. He received his Ph.D. in Computer Science from the University of Bath focusing on data aggregation routing in Wireless Sensor Networks. Saeid's research and teaching expertise centres on smart and adaptive computing and/or communication solutions to build collaborative/federated (sensory/feedback) systems in Internet of Things (IoT) applications and cloud environments. Saeid has two academic books and published more than 80 scholarly articles in reputed international journals and peer-reviewed conferences.

Ali Cheshmehzangi is a Full Professor and Head of the School of Architecture, Design and Planning (ADP) at the University of Queensland, where he leads the School and researches across the fields of architecture, design, and planning/urbanism. He has been in the World's top 2% field leader since 2022, recognised by Stanford University. He is among the top 30 global scholars in the urban sustainability research area. Ali has a wealth of experience, expertise, and a strong commitment to innovation and excellence in the fields of architecture, design, and planning, the three core disciplines of his School at UQ. With a career spanning over two decades, he has made significant contributions to the academic and professional communities, with a focus on sustainable and environmentally conscious design. So far, Ali has published over 500 journal papers, articles, conference papers, book chapters, and reports. He also has 28 other academic books, three of which have received awards at the national, provincial, and municipal levels. He also has received international awards and recognition for his research on urban resilience studies and sustainability research, as well as a Vice-Chancellor's award for his impactful contribution to higher education.

Embracing the Digital Twin Paradigm for Urban Sustainability

Ali Cheshmehzangi and Saeid Pourroostaei Ardakani

Abstract In the face of escalating urbanization, cities worldwide are seizing with the need to become more sustainable, efficient, and resilient. The insistent march of technology has gifted us with innovative tools that have the potential to transform urban landscapes. Among these, the concept of the Digital Twin has emerged as a groundbreaking paradigm, offering unprecedented capabilities for modelling, analyzing, and managing urban environments in real-time. This introduction chapter aims to provide a comprehensive overview of how Digital Twin Computing, as a technology, coupled with advancements in Geographic Information Systems (GIS), the Internet of Things (IoT), and Artificial Intelligence (AI), is reshaping urban planning and sustainability efforts. By leveraging these technologies, urban planners can gain deeper insights, optimize resource allocation, and enhance disaster response strategies. The chapter also provides a summary of the book's aim, objectives, and structure, setting the stage for an in-depth exploration of the transformative impact of Digital Twin technology on urban sustainability and resilience.

Keywords Digital twin computing · Urban intelligence · Digital revolution · Digital age · Digital technologies · Technological innovation · Urban sustainability · Smartness

1 Introduction: The Digital Twin Concept

At its core, a Digital Twin is a virtual replica of a physical entity, maintained through continuous data exchanges. This technology creates a dynamic and interactive model that mirrors the current state, historical performance, and potential future scenarios

---

A. Cheshmehzangi
School of Architecture, Design and Planning, The University of Queensland, Brisbane, Australia

S. P. Ardakani (✉)
School of Engineering and Physical Sciences , College of Health and Science, University of Lincoln, Lincoln, UK
e-mail: SPourroostaei@lincoln.ac.uk

S. Pourroostaei Ardakani and A. Cheshmehzangi (eds.), *Digital Twin Computing for Urban Intelligence*, Urban Sustainability,
https://doi.org/10.1007/978-981-97-8483-7_1

of its real-world counterpart [1–3]. Initially developed for industrial applications, Digital Twins are now finding extensive use in urban planning, where they can simulate everything from individual buildings to entire cities [4–7]. This capability allows for more informed decision-making, optimized resource management, and enhanced disaster preparedness [8]. Here, the focus is on Digital Twin Computing (DTC) where we see a growing potential in how it could be integrated in cities and urban environments and toward achieving urban intelligence in practice.

Digital Twin Computing (DTC) represents a convergence of multiple technological advancements including IoT, AI, cloud computing, and GIS [9–13]. This convergence is driving the transformation of urban environments into smart cities [14] characterized by high efficiency, sustainability, and improved quality of life for residents. The evolution of DTC in urban planning is indicative of a broader trend towards harnessing digital technologies [15] to address the complex challenges faced by modern cities.

Here, we summarise five domains where DTC could help improve the (urban) intelligence of cities, including—but not limited to—areas, such as "urban planning and design", "infrastructure and building management", "disaster preparedness and response", "smart urban services", and "civic engagement and participatory planning" (Fig. 1).

1.1 Enhancing Urban Planning and Design

In urban planning, the ability to simulate and analyze potential scenarios is crucial. Digital Twins enable planners to create detailed, data-rich models of urban environments [16–18]. These models can incorporate real-time data from IoT sensors, historical data, and predictive analytics to simulate various scenarios [19]. For example, planners can assess the impact of new infrastructure projects, optimize traffic flow, or evaluate the potential effects of climate change on urban areas. This capability is particularly valuable for developing sustainable urban designs [20] that minimize environmental impact and improve resilience to natural disasters.

Digital Twins can also facilitate more efficient resource management. By providing a comprehensive view of urban systems, these models enable city officials to optimize the allocation of resources such as energy, water, and transportation [21, 22]. For instance, a Digital Twin of a city's energy grid can help identify inefficiencies and recommend solutions to reduce energy consumption [23, 24]. Similarly, water management systems can be optimized to ensure a reliable supply and prevent wastage [25]. This level of optimization is essential for cities striving to become more sustainable and reduce their environmental footprint.

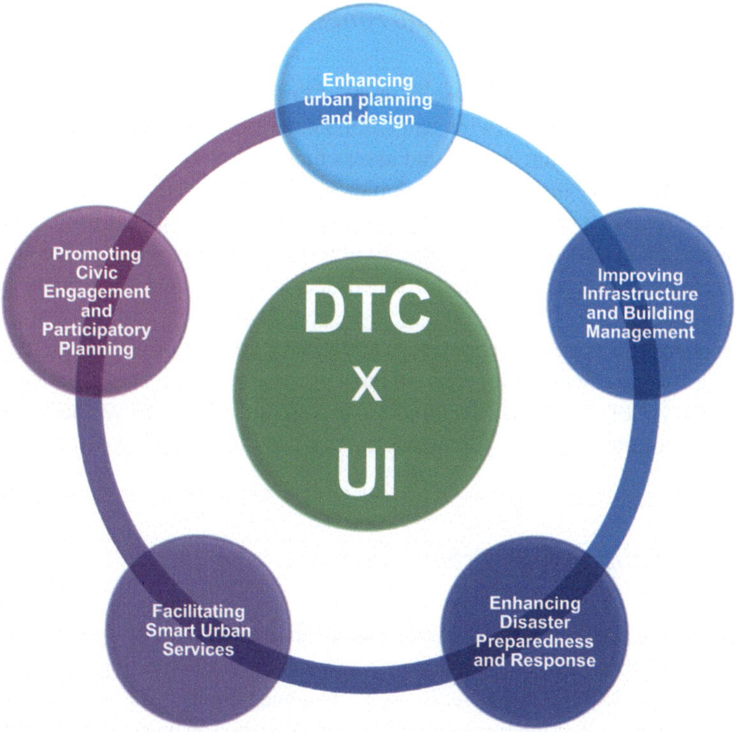

Fig. 1 Summary of five domains for DTC × UI integration

1.2 Improving Infrastructure and Building Management

The integration of DTC into infrastructure management offers significant benefits. Digital Twins can monitor the condition of infrastructure assets in real-time, predict maintenance needs, and extend the lifespan of these assets [26, 27]. For example, a Digital Twin of a bridge can continuously monitor stress levels, detect structural weaknesses, and predict when maintenance is required. This proactive approach to maintenance can prevent costly repairs and enhance the safety of infrastructure.

In the context of building management, Digital Twins can optimize the operation of HVAC systems, lighting, and other building services to reduce energy consumption and improve occupant comfort [28]. By integrating data from IoT sensors, building management systems can adjust settings in real-time based on occupancy levels, weather conditions, and other factors. This not only reduces energy costs but also contributes to the overall sustainability of the building.

1.3 Enhancing Disaster Preparedness and Response

Disaster preparedness and response is another critical area where DTC can make a significant impact. Digital Twins can simulate various disaster scenarios, such as earthquakes, floods, or hurricanes, and assess their potential impact on urban areas [29]. This enables city officials to develop more effective emergency response plans and allocate resources more efficiently [30]. During a disaster, real-time data from IoT sensors can be fed into the Digital Twin to provide a live view of the situation [31], helping responders to make informed decisions and coordinate their efforts more effectively.

For instance, in the event of a flood, a Digital Twin can provide real-time data on water levels, predict which areas are most at risk, and suggest evacuation routes [32]. This level of situational awareness can significantly improve the speed and effectiveness of the response, potentially saving lives and reducing damage.

1.4 Facilitating Smart Urban Services

The concept of smart cities encompasses a wide range of services designed to improve the quality of urban life. DTC can play a crucial role in delivering these services more effectively [33]. For example, smart waste management systems can use Digital Twins to optimize collection routes based on real-time data on waste levels in different areas [34]. Similarly, smart transportation systems can use Digital Twins to manage traffic flow, reduce congestion [35], and improve public transportation services.

Digital Twins can also support the development of smart utilities, such as electricity and water supply systems, by providing a comprehensive view of the entire network [36, 37]. This enables utilities to detect and address issues more quickly, improving reliability and reducing downtime. Additionally, smart grids can use Digital Twins to integrate renewable energy sources more effectively, balancing supply and demand to ensure a stable and sustainable energy supply.

1.5 Promoting Civic Engagement and Participatory Planning

One of the most exciting potentials of DTC lies in its ability to promote civic engagement and participatory planning [38]. Digital Twins can provide a platform for citizens to visualize proposed urban developments and provide feedback [39]. This level of transparency can foster greater trust and collaboration between city officials and residents, leading to more inclusive and effective urban planning.

For example, a Digital Twin of a proposed new park can allow residents to explore the design, suggest changes, and see how their input is incorporated into the final plan

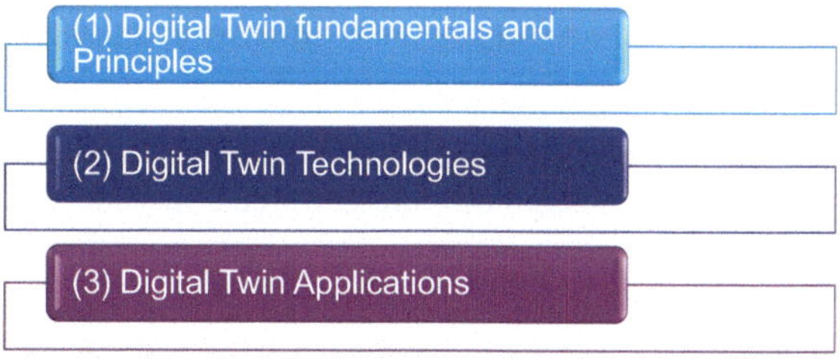

Fig. 2 Summary of three parts covered in the book

[40]. This collaborative approach not only improves the quality of the final design but also ensures that it meets the needs and preferences of the community.

2 Aim and Objectives of the Book

This edited volume focuses on new and ongoing discourses in interdisciplinary research and practice in urban system and smart city development pathways. It approaches digital twin fundamentals and principals including theoretical foundations, conceptualisations, strategies and services/patterns to define and adapt digital twin solutions for urban applications—mainly sustainability. To achieve this overarching aim, this book highlights promising case studies and outlines digital twin design models and system architecture by examining key digital twin deployment practices such as data analysis, decision making, and service automation in the line with intelligent urban planning. It also emphasises on DT technologies such as cloud computing, AI, IoTs, and smart virtualisation and outlines the key benefits of the DT solutions in urban applications—mainly control and planning.

3 The Structure: Reflection on Book Chapters

The book is divided into three parts of (1) Digital Twin fundamentals and Principles, (2) Digital Twin Technologies, and (3) Digital Twin Applications (see Fig. 2), comprising 11 review and case study chapter contributions from global scholars. It then ends with a concluding chapter that puts together a summary of direction for navigating the future with the nexus between digital twin computing and urban intelligence. Below are the reflective summaries of contributing chapter to this edited volume.

Chapter 2: Urban Rooftop Solar Energy Planning

The journey towards sustainable cities often begins with energy. Rooftop solar photovoltaic (PV) systems represent a vital component of urban renewable energy strategies. However, integrating these systems into the urban fabric is fraught with challenges related to spatial planning, resource optimization, and stakeholder engagement. Chapter "Towards Sustainable Urban Rooftop Solar Energy Planning Through Spatial Digital Twins Paradigm: A Systematic Literature Review" delves into the role of GIS-based methodologies in assessing the potential of rooftop solar energy. By systematically reviewing the literature, this chapter identifies the technical challenges, trends, and opportunities for enhancing urban solar energy adoption through advanced spatial planning and digital twin technologies. The integration of machine learning and three-dimensional (3D) mapping in GIS models is highlighted as a promising trend, pointing towards a future where spatial digital twins can support more effective solar energy strategies.

Chapter 3: Open-Source Solutions for Data-Scarce Regions

While data abundance is often taken for granted in developed regions, many parts of the world face significant data scarcity, hindering the development of sophisticated urban models. Chapter "Creating a Foundation for Spatial Digital Twins in Data-Scarce Regions Through Open-Source Solutions" explores how open-source solutions can bridge this gap, using Honiara, the capital city of the Solomon Islands, as a case study. By leveraging open-source GIS and cloud computing tools, a basic yet functional spatial digital twin of the city was created. This approach demonstrates that even data-scarce regions can benefit from digital twin technology, laying a foundation for future advancements in urban planning and management.

Chapter 4: Fusion of Digital Twin, IoT, and AI for Urban Intelligence

The fusion of Digital Twin technology with IoT and AI represents a transformative leap towards smarter cities. Chapter "Fusion of Digital Twin, Internet-of-Things and Artificial Intelligence for Urban Intelligence" investigates how these technologies can be integrated to enhance urban intelligence and decision-making processes. By collecting and analyzing vast amounts of data in real-time, cities can optimize resource allocation, improve infrastructure management, and foster greater civic engagement. AI-driven predictive analytics, in particular, hold the promise of making urban systems more responsive and efficient, thereby contributing to the overall sustainability and resilience of cities.

Chapter 5: Enhancing Disaster Management with Digital Twins and Generative AI

Disaster management is a critical aspect of urban resilience. Chapter "Building Resilient Smart Cities: The Role of Digital Twins and Generative AI in Disaster Management Strategy" focuses on the synergy between Digital Twin technology and Generative AI to improve disaster preparedness and response. Generative AI enhances digital twins by providing predictive capabilities and simulating complex scenarios. This combination allows for more accurate modeling of potential disasters, helping cities to develop and test effective mitigation strategies. The chapter underscores the importance of integrating these technologies to create more resilient urban environments capable of withstanding various emergencies.

Chapter 6: Optimizing Urban Heating with Geothermal Energy and Chemical Heat Pumps

Sustainable energy solutions extend beyond solar power. Chapter "Optimizing Urban Heating: Integrating Geothermal Energy and Chemical Heat Pumps for Digital Twin Simulations" examines the integration of geothermal energy and chemical heat pumps within Digital Twin simulations to optimize urban heating systems. By leveraging real-time data and advanced simulation capabilities, cities can improve the efficiency of these systems, reducing energy consumption and environmental impact. The chapter provides a comprehensive analysis of different thermal fluids and their performance, offering insights into the future of sustainable urban heating solutions.

Chapter 7: Optimizing Public Pedestrian Infrastructure

Public spaces such as parks, plazas, and streetscapes play a vital role in urban life. Chapter "Integrating Digital Twins in Urban Sustainability: A Framework for University Campus Applications" presents a novel approach to optimizing energy efficiency in these areas through multi-scale digital twin technology. By integrating real-time IoT sensor data and machine learning algorithms, cities can improve the usability and sustainability of public pedestrian spaces. The chapter outlines a comprehensive framework for data integration and simulation, offering a versatile tool for urban designers and planners.

Chapter 8: Consumer Energy Services and Digital Twin Applications

The role of Digital Twin technology in consumer energy services is another critical area of exploration. Chapter "A Data-Driven Multi-Scale Digital Twin Framework for Optimizing Energy Efficiency in Public Pedestrian Infrastructure" proposes a framework for monitoring and controlling energy usage in residential and urban areas. By integrating renewable energy sources and advanced technologies, digital twins can help optimize energy consumption, promoting sustainability and efficiency. This chapter highlights the transformative potential of digital twins in addressing the complex challenges of urban energy management.

Chapter 9: Framework for University Campus Applications

The scalability of Digital Twin technology is a significant advantage, enabling its application across various scales from individual buildings to entire cities. Chapter "Applications of Digital Twin Technology in Consumer Energy Services" introduces a systematic framework for developing digital twins for university campuses, which can be adapted to broader urban contexts. By integrating diverse data sources, this framework enhances facility management, asset lifecycle optimization, and overall campus sustainability. The methodologies outlined in this chapter aim to guide future large-scale digital twin projects, providing a model for urban sustainability efforts.

Chapter 10: Road Infrastructure Management in Uzbekistan

Transportation infrastructure is a critical component of urban planning. Chapter "Roads Infrastructure Digital Twin Case Study: Uzbekistan" explores the integration of Digital Twin technology into road infrastructure management in Uzbekistan. By incorporating real-time data, AI, and GIS, digital twins can enhance road maintenance, traffic management, and environmental resilience. The chapter provides a detailed case study, highlighting the potential benefits and socio-economic impacts of adopting digital twins for road infrastructure management.

Chapter 11: Urban Entrepreneurship and Tactical Urbanism

The final chapter, Chapter "Synergizing Digital Twin Computing with Urban Entrepreneurship and Tactical Urbanism: A Paradigm for Smart City Evolution", investigates the intersection of Digital Twin technology with urban entrepreneurship and tactical urbanism. By enabling comprehensive urban data analytics, collaborative design processes, and adaptive implementation, digital twins can support innovative urban interventions. This chapter presents a conceptual model for integrating digital twin computing into urban intelligence, emphasizing the importance of community participation and co-designed solutions in smart city development.

4 Concluding Remarks

The introduction of Digital Twin technology into urban planning marks a significant advancement in our ability to create sustainable, resilient, and intelligent cities. From optimizing energy systems and disaster management to enhancing public spaces and infrastructure, digital twins offer a versatile and powerful tool for urban planners, policymakers, and researchers. As cities continue to evolve, the integration of GIS, IoT, AI, and open-source solutions within the digital twin framework will be crucial in addressing the complex challenges of urban sustainability and resilience. This comprehensive exploration across various domains underscores the transformative potential of digital twins, paving the way for smarter and more sustainable urban futures.

The integration of Digital Twin Computing (DTC) into urban planning and management represents a significant step towards creating smarter, more sustainable cities. By providing a dynamic and interactive model of urban environments, Digital Twins enable more informed decision-making, optimized resource management, and enhanced disaster preparedness. The potential applications of DTC are vast, encompassing everything from infrastructure management and building optimization to smart services and civic engagement.

As cities continue to grow and face increasingly complex challenges, the adoption of Digital Twin technology will be essential for achieving urban intelligence in practice. By harnessing the power of digital technologies, cities can become more resilient, efficient, and liveable, paving the way for a sustainable future. The chapters that follow will delve deeper into specific applications of Digital Twin technology (DTC), exploring how it can be leveraged to address various aspects of urban sustainability and resilience. The focus is primarily on urban intelligence, where we believe it could be seen a great starting point for new opportunities and paradigm shifts in urbanism. Hence, the nexus between DTC and UI is the backbone of this book's aim and objectives.

References

1. Jeremiah SR, Yang LT, Park JH (2024) Digital twin-assisted resource allocation framework based on edge collaboration for vehicular edge computing. Futur Gener Comput Syst 150:243–254
2. Liu M, Fang S, Dong H, Xu C (2021) Review of digital twin about concepts, technologies, and industrial applications. J Manuf Syst 58:346–361
3. Singh M, Fuenmayor E, Hinchy EP, Qiao Y, Murray N, Devine D (2021) Digital twin: origin to future. Appl Syst Innov 4(2):36
4. Batty M (2024) Digital twins in city planning. Nat Comput Sci 4(3):192–199
5. Cheshmehzangi A, Batty M, Allam Z, Jones DS (2024) City Information Modelling: an Insight into a New Era for the built environment. In: City information modelling. Springer Nature Singapore, Singapore, pp 1–13
6. Ketzler B, Naserentin V, Latino F, Zangelidis C, Thuvander L, Logg A (2020) Digital twins for cities: a state of the art review. Built Environ 46(4):547–573
7. Shahzad M, Shafiq MT, Douglas D, Kassem M (2022) Digital twins in built environments: an investigation of the characteristics, applications, and challenges. Buildings 12(2):120
8. Ham Y, Kim J (2020) Participatory sensing and digital twin city: Updating virtual city models for enhanced risk-informed decision-making. J Manag Eng 36(3):04020005
9. Boje C, Kubicki S, Guerriero A, Rezgui Y, Zarli A (2022) Digital twins for the built environment. Buildings and semantics. CRC Press, New York, pp 179–199
10. Qian C, Liu X, Ripley C, Qian M, Liang F, Yu W (2022) Digital twin—cyber replica of physical things: architecture, applications and future research directions. Future Internet 14(2):64
11. Taherkhani R, Ashtari MA, Aziminezhad M (2024) Digital twin-enabled infrastructures: a bibliometric analysis-based review. J Infrastruct Syst 30(1):03123001
12. Wajid MS, Terashima-Marin H, Najafirad P, Pablos SEC, Wajid MA (2024) DTwin-TEC: an AI-based TEC district digital twin and emulating security events by leveraging knowledge graph. J Open Innov Technol Market Complex 10(2):100297
13. Zamari M (2023) A proposal for a Wildfire Digital Twin Framework through automatic extraction of remotely sensed data: the Italian case study of the Susa Valley. Doctoral dissertation, Politecnico di Torino
14. Cheshmehzangi A (2021) From transitions to transformation: a brief review of the potential impacts of COVID-19 on boosting digitization, digitalization, and systems thinking in the built environment. J Build Constr Plan Res 9(1):26–39
15. Cheshmehzangi A (2022) ICT for a sound business environment in cities: opportunities through digital technologies and digital transformations. In: ICT, cities, and reaching positive peace. Springer Nature Singapore, Singapore, pp 197–217
16. Faliagka E, Christopoulou E, Ringas D, Politi T, Kostis N, Leonardos D, Tranoris C, Antonopoulos CP, Denazis S, Voros N (2024) Trends in digital twin framework architectures for smart cities: a case study in smart mobility. Sensors 24(5):1665

17. Gil J (2020) City information modelling: A conceptual framework for research and practice in digital urban planning. Built Environ 46(4):501–527
18. Xu H, Omitaomu F, Sabri S, Li X, Song Y (2024) Leveraging generative AI for smart city digital twins: a survey on the autonomous generation of data, scenarios, 3D city models, and urban designs. arXiv preprint arXiv:2405.19464
19. Pourroostaei Ardakani S, Cheshmehzangi A (2023) Big data analytics and the future of smart transport and healthcare systems. In: Big data analytics for smart transport and healthcare systems. Springer Nature Singapore, Singapore, pp 175–184
20. Cheshmehzangi A, Dawodu A, Sharifi A (2021) Sustainable urbanism in China. Routledge, New York
21. Al-Sehrawy R, Kumar B, Watson R (2021) A digital twin uses classification system for urban planning & city infrastructure management. J Inf Technol Constr 26:832–362
22. Ford DN, Wolf CM (2020) Smart cities with digital twin systems for disaster management. J Manag Eng 36(4):04020027
23. Francisco A, Mohammadi N, Taylor JE (2020) Smart city digital twin–enabled energy management: toward real-time urban building energy benchmarking. J Manag Eng 36(2):04019045
24. Jafari M, Kavousi-Fard A, Chen T, Karimi M (2023) A review on digital twin technology in smart grid, transportation system and smart city: challenges and future. IEEE Access 11:17471–17484
25. Ghaith M, Yosri A, El-Dakhakhni W (2022) Synchronization-enhanced deep learning early flood risk predictions: the core of data-driven city digital twins for climate resilience planning. Water 14(22):3619
26. Khallaf R, Khallaf L, Anumba CJ, Madubuike OC (2022) Review of digital twins for constructed facilities. Buildings 12(11):2029
27. Liu Z, Blasch E, Liao M, Yang C, Tsukada K, Meyendorf N (2023) Digital twin for predictive maintenance. NDE 4.0 Predictive Maintenance, Commun Energy Syst Digit Transf NDE 12489:27–37
28. Bortolini R, Rodrigues R, Alavi H, Vecchia LFD, Forcada N (2022) Digital twins' applications for building energy efficiency: a review. Energies 15(19):7002
29. Ariyachandra MMF, Wedawatta G (2023) Digital twin smart cities for disaster risk management: a review of evolving concepts. Sustainability 15(15):11910
30. Cheshmehzangi A, Sedrez M, Flynn A (2024) Rethinking stormwater management through sustainable urban design. Springer Nature Singapore, Singapore
31. Alibrandi U (2022) Risk-informed digital twin of buildings and infrastructures for sustainable and resilient urban communities. ASCE-ASME J Risk Uncertain Eng Syst Part A: Civ Eng 8(3):04022032
32. Henriksen HJ, Schneider R, Koch J, Ondracek M, Troldborg L, Seidenfaden IK, Stisen S (2022) A new digital twin for climate change adaptation, water management, and disaster risk reduction (HIP digital twin). Water 15(1):25
33. Mylonas G, Kalogeras A, Kalogeras G, Anagnostopoulos C, Alexakos C, Muñoz L (2021) Digital twins from smart manufacturing to smart cities: a survey. IEEE Access 9:143222–143249
34. Li N, Padwal HH (2020) The importance of public support in the implementation of green transportation in smart cities using smart vehicle bicycle communication transport. Electron Libr 38(5/6):997–1011
35. Cheshmehzangi A, Pourroostaei Ardakani S (2021) Urban traffic optimization based on modeling analysis of sector-based time variable: the case of simulated Ningbo, China. Front Sustain Cities 3:629940
36. Ramos HM, Kuriqi A, Besharat M, Creaco E, Tasca E, Coronado-Hernández OE, Pienika R, Iglesias-Rey P (2023) Smart water grids and digital twin for the management of system efficiency in water distribution networks. Water 15(6):1129
37. Zekri S, Jabeur N, Gharrad H (2022) Smart water management using intelligent digital twins. Comput Inf 41(1):135–153

38. Chen W, Cheshmehzangi A, Mangi E, Heath T, Ye C, Wang L (2022) An analysis of residents' social profiles influencing their participation in community micro-regeneration projects in China: a case study of Yongtai Community, Guangzhou. Land 11(6):790
39. Peldon D, Banihashemi S, LeNguyen K, Derrible S (2024) Navigating urban complexity: the transformative role of digital twins in smart city development. Sustain Cities Soc 111:105583
40. Dani AAH, Supangkat SH, Lubis FF, Nugraha IGBB, Kinanda R, Rizkia I (2023) Development of a smart city platform based on digital twin technology for monitoring and supporting decision-making. Sustainability 15(18):14002

Digital Twin Fundamentals and Principles

Towards Sustainable Urban Rooftop Solar Energy Planning Through Spatial Digital Twins Paradigm: A Systematic Literature Review

Athenee Teofilo and Qian Sun

Abstract With the growing need for sustainable urban energy solutions, rooftop solar photovoltaic (PV) systems can play a pivotal role. However, the effective integration of solar energy into urban landscapes faces challenges in spatial planning, resource optimisation, and stakeholder engagement. This literature review addresses the existing gaps by systematically evaluating Geographic Information System (GIS)-based methodologies for PV potential assessment and rooftop solar energy planning. It aims to provide the knowledge necessary for supporting efficient and sustainable solar energy adoption in cities. This chapter involved a systematic review and synthesised the diverse body of knowledge regarding city-scale rooftop solar energy planning. The goal was to identify the technical challenges, highlight the trends, and reveal opportunities for future research initiatives with respect to the evolving needs of sustainable energy planning. Employing a dual-method systematic approach, this review utilised both quantitative and qualitative analysis techniques. Through the quantitative survey, a generic corpus (805 studies) of Scientometrics using CiteSpace was built, ensuring a broad representation of the existing literature. In parallel, a qualitative survey, evaluating literature associated with solar rooftop PV assessment using GIS, modelling techniques, planning implications, and recommendations, led to a critical corpus (76 studies) using the Preferred Reporting Items for Systematic Reviews (PRISMA) method. Spanning the past two decades and drawing from databases of Scopus and Web of Science (WoS), the synthesised corpora formed the foundation for a comprehensive and informed evaluation. Additionally, analytical assessment criteria were developed to provide insights into future research agendas and framework. The literature synthesis unveiled the challenges within urban energy planning, covering issues of data accuracy, model validation, and imperative stakeholder engagement. Emerging trends in GIS models, especially the integration of machine learning and three-dimensional (3D) mapping, emphasise the

A. Teofilo (✉) · Q. Sun
School of Science, RMIT University, Melbourne, Australia
e-mail: anee.teofilo@rmit.edu.au

Q. Sun
e-mail: chayn.sun@rmit.edu.au

S. Pourroostaei Ardakani and A. Cheshmehzangi (eds.), *Digital Twin Computing for Urban Intelligence*, Urban Sustainability,
https://doi.org/10.1007/978-981-97-8483-7_2

dynamic nature of the urban environment. Furthermore, applications of GIS in real-world urban contexts provide spatially explicit insights for future implementations. Finally, this literature review proposed a research agenda for advancing GIS-based rooftop solar energy planning, with a particular focus on developing a spatial digital twin environment incorporating three-dimensional (3D) models, real-time data integration, and decision support systems tailored for city-scale applications. Spatial digital twin technologies enable modelling of different energy production scenarios. Such an integrated approach will underpin effective strategies to accommodate the complexities within urban environments, thus providing a roadmap for researchers, urban planners, and policymakers to collaboratively develop sustainable solutions for rooftop solar energy adoption at a city scale.

Keywords Rooftop solar photovoltaic (PV) assessment · Spatial planning · Energy sustainability

1 Introduction

Urban areas face significant challenges in achieving sustainable energy management, particularly in harnessing the potential of rooftop solar energy [3, 53, 112]. As cities expand and populations grow, there is an increasing demand for energy, exacerbating environmental concerns [42, 45, 98]. This indicates emphasises the need for renewable energy sources [33, 43, 45].

Rooftop solar presents a promising solution, however, its effective utilisation requires addressing technical gaps in spatial representation, modelling, and spatial planning. Spatial representation in urban contexts involves accurately capturing the complex geometries of buildings, their orientations, and surrounding obstructions [96], which significantly impacts solar energy availability. Traditional methods such as the constant-value approach and manual selection methods often fall short in capturing these intricate components, leading to inaccuracies in solar potential assessments [68]. However, advancements in Geographic Information Systems (GIS) offer opportunities to improve spatial representation by integrating detailed geographical data with building information models (BIM) and LiDAR scans [60]. Replicating this approach in solar rooftop PV studies enables a more precise solar irradiance modelling.

Modelling rooftop solar potential entails predicting solar radiation levels on rooftops based on factors including geographical location, building characteristics, and environmental conditions [28, 35, 64]. Existing 2D models may lack granularity and can fail to account for localised effects such as shading from nearby structures or vegetation [29, 75, 85]. Incorporating advanced GIS techniques, such as viewshed analysis [62, 83] and 3D solar irradiance modelling [87, 110], can enhance the accuracy of solar potential assessments by accounting these spatial complexities.

Spatial planning for rooftop solar energy involves optimising the placement of solar installations within urban landscapes while considering regulatory constraints, land use patterns, and community preferences [58]. Traditional planning approaches often overlook spatial nuances [47] and fail to engage stakeholders effectively [33] in the decision-making process. By leveraging GIS-based spatial decision support systems (SDSS) [94, 99], planners can visualise and analyse the impacts of different solar deployment scenarios, facilitate stakeholder participation through interactive mapping tools, and streamline the planning process.

This chapter seeks to address technical gaps in spatial representation, modelling, and spatial planning for rooftop solar energy in urban areas. By synthesising existing research and identifying opportunities for improvement, this study aims to propose an innovative approach that utilises GIS technologies with the ultimate goal of supporting sustainable rooftop solar energy planning. To achieve this aim, the following objectives are established to:

- Undertake a quantitative review of the existing practises, findings, and emerging trends in rooftop solar energy for the development of a scientometric generic corpus;
- Conduct a qualitative review using analytical assessment criteria from the critical corpus to obtain insight into GIS-based studies of solar rooftop PV potential;
- Synthesise the research gaps and identify opportunities to enhance data representation, model validation, and stakeholder engagement;
- Propose a conceptual framework for a comprehensive modelling of city-scale rooftop solar energy potential.

2 Methodology

2.1 Overall Workflow

A dual-method approach comprising both quantitative and qualitative analysis techniques has been used in this study. CiteSpace, developed by Chen [20], has been used for the quantitative survey to obtain the scientometric generic corpus, while the Preferred Reporting Items for Systematic Reviews (PRISMA) method, introduced by Moher et al. [71], was employed for the qualitative survey of the study to develop the critical corpus. Moreover, to extract insights from the critical corpus, an analytical assessment criterion was applied. This was used to derive the future research agenda and present the prospective conceptual framework for a sustainable urban rooftop PV energy planning. The overall workflow is illustrated in Fig. 1.

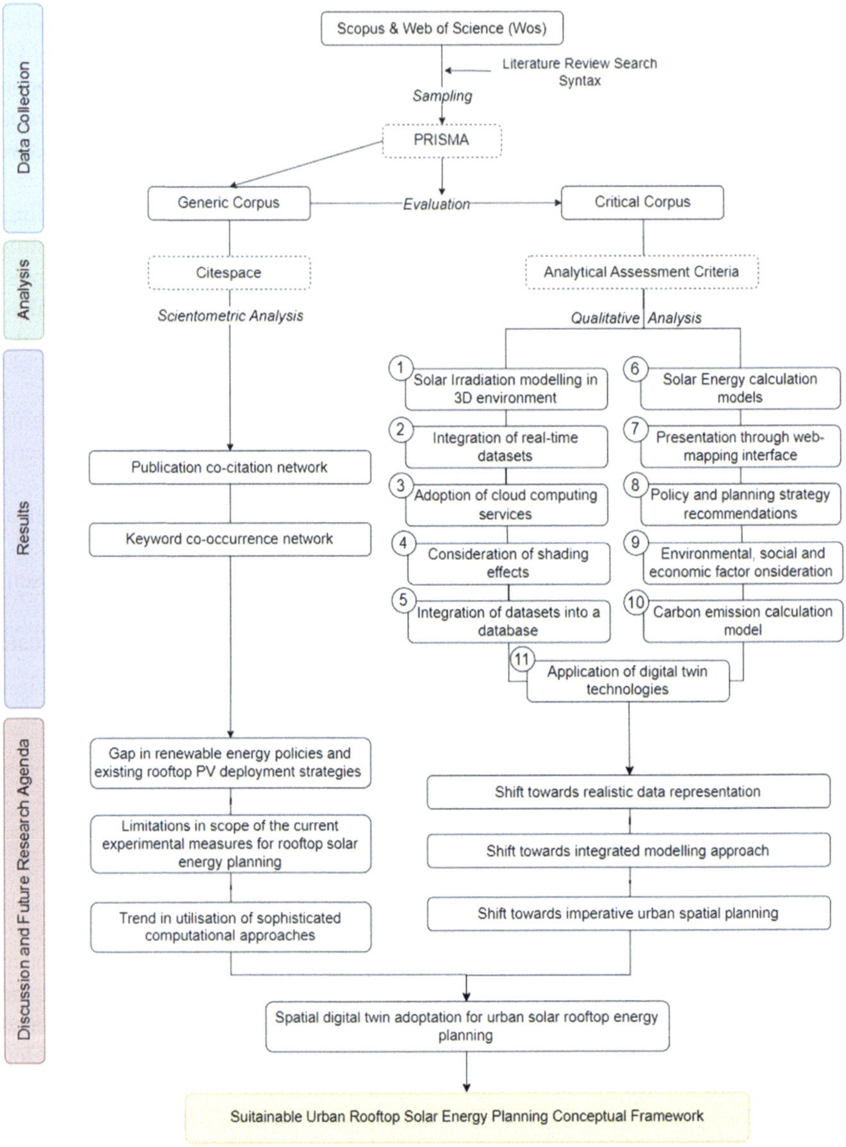

Fig. 1 Literature review outline showing the overall literature review workflow

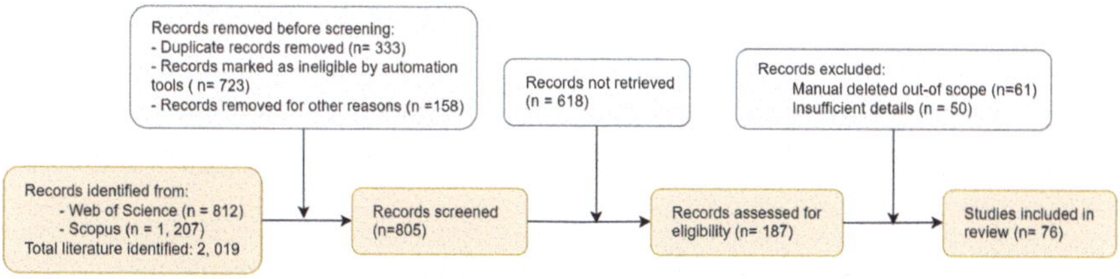

Fig. 2 Development of the critical corpus using the PRISMA workflow

2.2 Data Collection

2.2.1 Building a Generic Corpus Through Literature Database Extraction

To obtain the generic corpus, this chapter is guided by the following criteria for the selection of literature from academic databases: (a) Covers all papers until August 11, 2023; (b) Targets the Urban Solar Rooftop PV Planning; (c) Written in English; (d) Contains one or more search criteria in either the Title, Abstract, or Keywords; (e) Journal paper. Both Web of Science and Scopus databases have been selected for this study. The syntax *"("city" OR "urban" OR "town" OR "metropolitan" OR "district") AND ("pv" OR "photovoltaic" OR "solar cell") AND ("sun" OR "solar") AND (roof*)"* was used. The search criteria were '*Title, Abstract, or Keywords*' for Scopus and '*Topic**' for WoS.

2.2.2 Building a Critical Corpus Using PRISMA Method

To obtain the critical corpus of the study, literature included from the generic corpus has been screened, assessed and analysed using the PRISMA method. Figure 2 shows the progression of in building the critical corpus.

2.3 Data Analysis

2.3.1 Quantitative Assessment in CiteSpace

The quantitative assessment involved the scientometric analysis of the generic corpus in CiteSpace with the objective of gaining insights into the current publication co-citation network and keyword co-occurrence network. The publication co-citation network presents the relationship between literature and its frequency in being cited on other academic papers, while the keyword co-occurrence network displays the connection of keywords within the literature.

For the publication co-citation network, nodes represent the authors, while the links represent co-citation relationships between two literatures. Simultaneously, for the keyword co-occurrence network, the links and nodes represent the relationship between two words and keywords, respectively.

2.3.2 Qualitative Assessment Using Analytical Assessment Criteria

In the qualitative assessment, the critical corpus has been inspected using analytical assessment criteria. Table 1 presents the analytical assessment criteria applied to the literature for the identification of gaps, recommendations, and potential research agendas.

3 Results

3.1 Quantitative Analysis Results

The networks generated from the scientometric analysis provided insights into the strength of connections and betweenness centrality between literature. The intensity or thickness of the line connecting two nodes indicates the strength of the link between two literatures. Betweenness centrality, reflecting the importance of a node within the network, is expressed as a value that ranges from 0 to 1.

3.1.1 Publication Co-citation Networks

For this chapter, the resulting publication co-citation network involved 817 nodes and 13, 516 links. This analysis enabled the identification of both highly cited authors and most influential authors within the literature landscape, as illustrated in Fig. 3b. [80] was found to have highest centrality with a co-citation value of 3, followed by Ordenes et al. [46] as the second highest centrality score with the most co-citation value. Subsequently, Nann and Emery [73], Wiginton et al. [103], and Brinkworth and Sandberg [16] were identified to be the third to fifth highest centrality score with a co-citation value of 2, 3, and 57, respectively.

Furthermore, the dominant themes in the research landscape have been determined, represented by the different clusters in Fig. 3a. A total of 19 clusters have been identified with the largest group centralised around the "#0 rooftop pv" cluster, which connects most of the other clusters. Additionally, there are extended connections to clusters "#10 non-ideal position", "#8 energy analysis", and "#4 experimental measurements", with the latter being the only cluster that has a continued link.

Table 1 Analytical assessment criteria

#	Question	Justification
1	Was solar irradiation modelled in a 3D environment?	Solar irradiation needs to be modelled in a 3D environment because it accounts for vertical spatial variations, roof orientation, and urban context, which influences the variation of solar irradiation values depending on the sun's position
2	Was any real-time dataset used in calculating solar energy?	Real-time datasets need to be integrated into the solar energy assessments to increase accuracy and improve spatial calculations, as natural factors such as solar irradiance, meteorological condition, and shading effects are time-dynamic and stochastic in nature
3	Was cloud computing used in the assessment of solar energy potential?	Cloud computing offers several benefits for solar energy planning assessments, including its capacity to handle large volume of data, conducting complex spatial analysis and providing a collaborative platform for timely decisions [26]
4	Was shading effects considered in the distribution of solar energy potential?	Shading needs to be considered in solar urban PV planning due to its impacts on urban solar energy yield [13]. By considering shading patterns, the identification of sites with minimal energy production potential due to shading can be avoided, while also highlighting sites with optimal production potential
5	Were the datasets integrated into a database?	Datasets play a beneficial role in solar energy analysis by streamlined data access, organisation, and management. This contributes to improved data efficiency, reliability, and consistency, particularly considering that solar energy analysis often involves a variety of spatial datasets sourced from diverse data authorities with differing formats and availability
6	Was energy calculation model used in the solar potential assessment?	Solar energy calculation model is important for a systematic quantification of energy generation especially on urban environments because it allows inclusion and flexibility in the simulation of various spatial components in determining solar rooftop energy potential
7	Was a web-mapping interface used to present the solar energy planning results?	Web-mapping interfaces offer significant advantages as they facilitate the effective dissemination of information, thereby fostering awareness and trust in the decision-making process for implementing urban solar energy planning initiatives
8	Were there solar energy policy or planning strategy recommendations?	Urban solar rooftop PV studies need to present policy suggestions and recommend strategies to assist stakeholders in the best approach to optimise and maximise the production of renewable energies to achieve both economic and environmental goals
9	Was there environment, social and economic factors considered in the solar rooftop deployment assessment?	Taking into account environmental, social, and economic factors promote sustainable developments that ensures equal opportunity for communities and the environment

(continued)

Table 1 (continued)

#	Question	Justification
10	Was there a calculation model to track the net-zero emission goal?	A calculation model for environment impact tracking is essential as it acts as a support tool in the evaluation of urban solar energy performance, identifying opportunities for improvement and guiding future tasks
11	Was digital twin technology used in the study?	Adapting spatial digital twin technologies for planning solar rooftop PV deployments is important as it enables modelling of different energy production scenarios, contributing to more resilient planning efforts

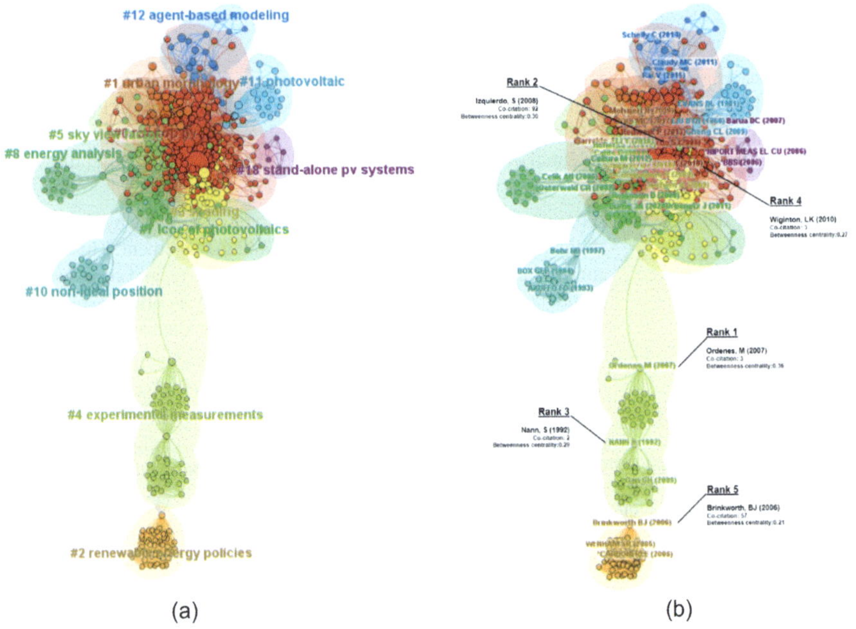

Fig. 3 Publication co-citation network showing **a** the different clusters and **b** the different authors and their corresponding centrality score ranking. *Source* Generated using citespace, v6.3.R2 advanced

3.1.2 Keyword Co-citation Networks

For the keyword co-occurrence network, 758 nodes and 19,973 links were observed. It identified the trend or the major focus areas of the topic over a period and their reflective prominence, indicated by their frequency. Figure 4 presents the keyword co-occurrence network with the rankings according to centrality score and corresponding co-occurrence frequency counts.

The keyword "solar energy" has the highest rank with a co-occurrence value of 0.23, followed by the clusters "photovoltaic system" and "photovoltaic cells" in ranks

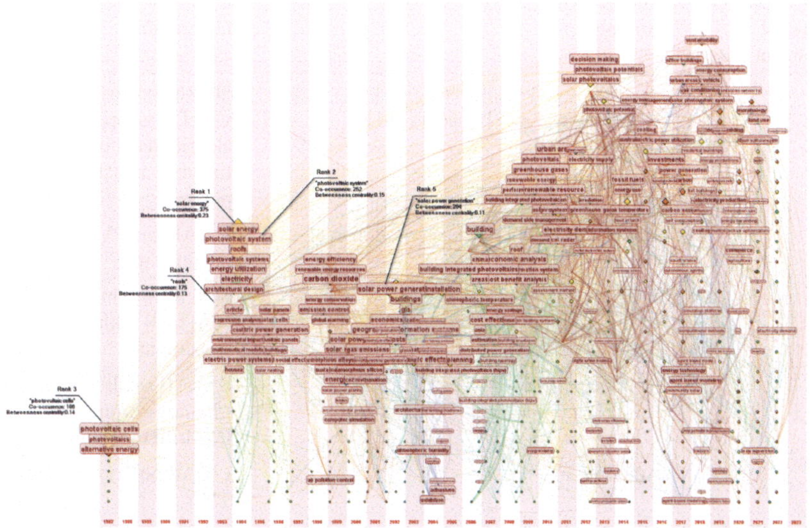

Fig. 4 Keyword co-occurrence network in timeline view. *Source* Generated using citespace, v6.3.R2 advanced

second and third with the co-occurrence values of 252 and 160, respectively. Then keywords "roofs" and "solar power generation" hold the fourth and fifth rank with co-occurrence values of 175 and 294, respectively.

The study spans from 1987 to 2023, revealing significant shifts in the occurrences of keywords. In the first half of the research landscape, aside from the top five keywords mentioned previously, terms such as "alternative energy", "environmental impact", "energy efficiency", "environmental protection", and "carbon dioxide" prevalent. Over time, new keywords gained prominence, including "solar power generation", "installation", "building integrated photovoltaics", "building", "economic analysis", "renewable energy", "greenhouse gases", "urban area", "decision making", "investments", "sustainability", and "energy consumption". Notably, emerging words such as "computational methods", "demand side management", "digital surface models", "energy modelling", "gis", "performance rate", "image segmentation", and "simulation platform" have also been identified.

3.2 *Qualitative Analysis Results*

The evaluation of the critical corpus revealed urban rooftop planning concepts have predominantly utilised a multi-factor approach for solar rooftop PV assessment. The results for the analytical assessment are presented in Table 2.

Table 2 Analytical assessment result

Analytical assessment criteria	Authors	Description
1	An et al. [5], Boulahia et al. [15], Kaleshwarwar and Bahadure [51]	3D Rhinoceros software with Grasshopper algorithm editor, Ladybug and Honeybee plugin
	Borfecchia et al. [14], Fuster-Palop et al. [31], Özdemir et al. [81]	LiDAR and PVGIS radiation data
	Brito et al. [17]	SOL algorithm in MATLAB with DSM and Hyperpoints
	Buffat et al. [18]	SARAH satellite solar irradiance profile and DSM/LiDAR point cloud
	Adjiski et al. [1]	LiDAR point cloud and TMY
	Han et al. [37]	TMY3 with LOD1 and LOD2 models
	Desthieux et al. [26]	LiDAR, 2D, and 3D cadastral with METENORM
	Ninsawat and Hossain [78]	Extruded 3D Models with Meteorological Department Data
	Lukac et al. [61]	LiDAR with Pyranometer Data
	Nguyen et al. [77]	Reflected on the paper but elaborated the process
	Chow et al. [21]	HOMER with DSM
	Salim et al. [89], Stack and Narine [93], Yesilmaden and Dogru [107]	DSM with ArcGIS Area Solar Radiation
	Zhang et al. [109]	DSM with PVSYST
	Davybida et al. [24]	DSM and SEBE model
	Yan et al. [105]	DSM/LiDAR and adopted model from another research
	Huang et al. [40]	DSM/LiDAR and SHORTWAVE-C model
	Verso et al. [100]	DSM with Meteosat
	Wong et al. [104]	DSM with observatory climatic data
	Gawley and McKenzie [34]	Used this for validation purposes

(continued)

Table 2 (continued)

Analytical assessment criteria	Authors	Description
	Thebault et al. [97]	Developed 3D-GIS-based irradiation model
	Jakubiec and Reinhart [49]	3D LiDAR with Daysim Simulation backward-raytracing engine
	Oh and Park [79]	DSM/DEM/Imagery and TMY
	Hubinsky et al. [41]	r.sun with LOD3
	Kozlovas et al. [57]	Photogrammetric Model and independent Solar Potential Calculator
2	Fuster-Palop et al. [31]	API for energy consumption for the computation of self-sufficiency
	Kozlovas et al. [57]	Automated smart meters of power data for consumption rates
3	Buffat et al. [18], Desthieux et al. [26]	Virtual Machine For the computation tasks
4	Adjiski et al. [1]	UMEP daily Shadow Pattern
	Alvarado et al. [4], Zubair et al. [113]	Exclusion criteria
	An et al. [5], Zhang et al. [109]	Elimination of shaded area
	Ninsawat and Hossain [78]	Used Sun Shadow Volume tool in the 3D Analyst extension
	Ayodele et al. [9], Bergamasco and Asinari [12], Cuesta-Fernández et al. [23]; Dehwah et al. [25]; Doorga et al. [27], Izquierdo et al. [46], Nguyen and Pearce [76], Verso et al. [100]	Estimated coefficient for shading effects
	Bazan et al. [11]; Mishra et al. [70]; Singh and Banerjee [91], Yang et al. [106]	Applied a correction factor/formula
	Yan et al. [105]	Modelled by 3D parallelograms
	Wang et al. [101]	Shading formula
	Borfecchia et al. [14], Gagnon et al. [32], Massano et al. [66], Nguyen et al. [77], Roth and Tilk [88], Salim et al. [89]	Included but no further explanation of integration

(continued)

Table 2 (continued)

Analytical assessment criteria	Authors	Description
	Boulahia et al. [15]	Ladybug plugin for the shadow elimination spacing
	Brito et al. [17], Gooding et al. [36], Sun et al. [95], Thebault et al. [97]	Included on the solar radiation module used
	Buffat et al. [18]	Used Sun Positioning algorithm
	Chow et al. [21]	Included in the Ecotect Analysis
	Assouline et al. [7, 8], Cieślak and Eźlakowski [22], Han et al. [37], Hong et al. [38], Kouhestani et al. [56], Margolis et al. [64] Stack and Narine [93]	Used Hillshade Tool
	Davybida et al. [24]	Used SEBE Model
	Desthieux et al. [26]	Shadow casting routine in SVF
	Fuster-Palop et al. [31]	Adopted a shadowing model
	Gawley and McKenzie [34]	Included ESRI solar calculation principles and Khanna technical workflows
	Huang et al. [40]	Used SHORTWAVE-C model
	Jakubiec and Reinhart [49]	Included in the Daysim simulation
	Jo et al. [50]	Used buffer of 150 ft. to eliminate and account the potential rooftop shading
	Karteris et al. [52]	Google SketchUp 3D Modelling
	Khan and Arsalan [53, 54], McIntyre [67]	Used reduction factor
	Khan et al. [55]	Applied averaged value from PYSyst
	Kozlovas et al. [57]	Used suitability factor
	Lukac et al. [61], Mikovits et al. [69]	Included on the estimation or solar radiation model
	Mangiante et al. [63]	Extracted from DSM
	Muhammed et al. [72], Wong et al. [104]	Shadow detection algorithm

(continued)

Table 2 (continued)

Analytical assessment criteria	Authors	Description
	Oh and Park [79]	VIEWMAP-based calculation
	Özdemir et al. [81]	Incorporated manually by estimated system losses in PVGIS
	Palmer et al. [82]	Hillshading from SAGA Software
	Ramirez Camargo et al. [86]	r.sun/r.horizon module
	Santos et al. [90]	Used PVGIS values
	Zalamea-León et al. [108]	Generated with BIM
5	Özdemir et al. [81]	PV potential information about each roof surface
	Yan et al. [105]	For data augmentation and transfer learning strategies
6	Adjiski et al. [1], Akrofi and Okitasari [2], Alvarado et al. [4], Assouline et al. [7, 8], Ayodele et al. [9], Baghani [10], Boulahia et al. [15], Chen et al. [19], Cieślak and Eźlakowski [22], Doorga et al. [27], Han et al. [37], Hong et al. [38], Jakubiec and Reinhart [49], Kaleshwarwar and Bahadure [51], Karteris et al. [52], Khan and Arsalan [53, 54], Khan et al. [55], Kouhestani et al. [56], Lukac et al. [61], Mangiante et al. [63], McIntyre [67], Neves et al. [74], Quiquerez et al. [84], Singh and Banerjee [91], Stack and Narine [93], Verso et al. [100], Wang et al. [101], Widodo et al. [102], Wong et al. [104], Zhong et al. [111], Zubair et al. [113]	Used Mathematical Model
	Asif et al. [6], Yang et al. [106]	Used a PVSyst Software
	Chow et al. [21]	HOMER Simulation Tool
	Cuesta-Fernández et al. [23]	Adopted a Peak Power formula
	Dehwah et al. [25]	PVSOL Software (Valentin), ARCGIS
	Desthieux et al. [26]	Hay Model
	Gawley and McKenzie [34]	Electricity production model Standard assessment Procedure (SAP) from BRE
	Gooding et al. [36], Mikovits et al. [69], Özdemir et al. [81]	Used PVGIS

(continued)

Table 2 (continued)

Analytical assessment criteria	Authors	Description
	Hubinsky et al. [41]	r.sun in QGIS
	Jo et al. [50]	Area Solar Radiation tool in ArcGIS and SAM
	Kozlovas et al. [57]	Used Solar Potential Calculator
	Margolis et al. [64], Zalamea-Leon et al. [108], Zhang et al. [109]	System Advisor Model (SAM)
	Massano et al. [66]	PV-Sim
	Mishra et al. [70], Roth and Tilk [88], Salim et al. [89], Sun et al. [95]	Area Solar Radiation tool in ArcGIS
	Muhammed et al. [72]	ArcGIS, PVWatts, PVGIS
	Nguyen et al. [77]	Reflected on the Workflow but was not elaborated
	Oh and Park [79]	Developed a tool ''SNU Solar", which is based in PVWatts, that includes calculation from SAM
	Ramirez Camargo et al. [86]	Used a mathematical model with r.sun
	Santos et al. [90]	Solar radiation using ArcGIS and PVGIS
	Thebault et al. [97], Yan et al. [105], Yesilmaden and Dogru [107]	Not presented
7	Alvarado et al. [4]	Used Openlayers with WMS file formats
	Brito et al. [17]	WebGIS viewer
	Desthieux et al. [26]	Updated Geneva Solar Cadastre
	Jakubiec and Reinhart [49]	Online mapping Application
	Massano et al. [66]	Developed a platform design
	Oh and Park [79]	Provided a graphical user interface (GUI)
8	Adjiski et al. [1], Baghani [10], Khan et al. [55], Kouhestani et al. [56], Muhammed et al. [72], Widodo et al. [102]	Empirical finding of a case study

(continued)

Table 2 (continued)

Analytical assessment criteria	Authors	Description
	Akrofi and Okitasari [2], An et al. [5], Ayodele et al. [9], Bazan et al. [11], Cieślak and Eźlakowski [22], Davybida et al. [24], Fuster-Palop et al. [31], Gagnon et al. [32], Han et al. [37], Hubinsky et al. [41], Jo et al. [50], Margolis et al. [64], Massano et al. [66], McIntyre [67], Mikovits et al. [69], Neves et al. [74], Singh and Banerjee [91], Stack and Narine [93], Thebault et al. [97], Wang et al. [101], Zalamea-León et al. [108], Zhang et al. [109], Zhong et al. [111], Zubair et al. [113]	Planning strategy recommendation
	Asif et al. [6], Boulahia et al. [15], Cuesta-Fernández et al. [23], Dehwah et al. [25], Gooding et al. [36], Kozlovas et al. [57], Mangiante et al. [63], Mishra et al. [70], Salim et al. [89], Santos et al. [90], Sun et al. [95]	Policy and planning strategy recommendation
	Gawley and McKenzie [34], Khan and Arsalan [53, 54], Yang et al. [106]	Policy recommendation
9	Akrofi and Okitasari [2], Kozlovas et al. [57], Zubair et al. [113]	Economic (Levelized Cost of Energy OR LCOE)
	Akrofi and Okitasari [2]	Social (Density)
	Alvarado et al. [4]	Environmental (Energy Consumption Offset)
	An et al. [5], Bazan et al. [11]	Environmental (Life cycle analysis or LCA)
	Asif et al. [6]	Economic (Benefit cost ratio)
	Asif et al. [6], Chen et al. [19], Fuster-Palop et al. [31], Gawley and McKenzie [34], Gooding et al. [36], Neves et al. [74], Ninsawat and Hossain [78], Sun et al. [95], Thebault et al. [97], Zubair et al. [113]	Economic (Payback Period)
	Chen et al. [19], Kouhestani et al. [56], Sun et al. [95], Wang et al. [101]	Economic (Net present value or NPV)
	Chen et al. [19], Fuster-Palop et al. [31], Mangiante et al. [63], Neves et al. [74]	Economic (Internal rate of return or IRR)
	Chow et al. [21]	Economic (Time-of-Use savings)
	Chen et al. [19], Gawley and McKenzie [34], Massano et al. [66], Muhammed et al. [72], Neves et al. [74], Salim et al. [89], Thebault et al. [97], Wang et al. [101]	Environmental (GHG emission/Carbon Emission Saving)

(continued)

Table 2 (continued)

Analytical assessment criteria	Authors	Description
	Fuster-Palop et al. [31], Neves et al. [74]	Economic (Self-sufficiency Rate)
	Fuster-Palop et al. [31], Neves et al. [74]	Economic (Self-consumption Rate)
	Lukac et al. [61]	Economic (Return of Investment or ROI)
	Neves et al. [74]	Economic (Surplus Rate)
	Stack and Narine [93]	Economic (Initial Investment Cost)
	Thebault et al. [97]	Social (Heritage Constraints)
10	Asif et al. [6]	Used RETScreen
	Bazan et al. [11]	LCA methodology with ISO 14040/14044 framework and SimaPro software
	Massano et al. [66], Muhammed et al. [72], Neves et al. [74], Sun et al. [95], Thebault et al. [97]	Mathematical model
	Gawley and McKenzie [34], Salim et al. [89], Wang et al. [101]	Carbon intensity (CI) conversion rate
	Widodo et al. [102]	CO_2 Reduction coefficient
	Chow et al. [21]	Results are shown but the process was not mentioned

3.2.1 Modelling of Solar Irradiation

Among the 76 reviewed literatures, the modelling of solar irradiation in a 3D environment constitutes 39.47% as the chosen approach for obtaining this value. The most utilised method involves the use of LiDAR point cloud data to reconstruct the study area in different 3D levels of detail. Furthermore, 3D modelling software such as Rhinoceros software with the Grasshopper algorithm editor, in conjunction with the Ladybug and Honeybee plugin, has been commonly adopted for this purpose.

3.2.2 Utilisation of Real-Time Dataset

The utilisation of real-time datasets was found to be scarce across the reviewed papers, with only 2.63% of the total papers utilising this data with temporal accuracy. The use of live data of energy consumption from buildings represented the sole instance were real-time datasets within solar energy studies.

3.2.3 Adoption of Cloud Computing

The integration of cloud computing services for solar energy analysis has been found to be uncommon, with its adoption observed only in about 2.63% of the total reviewed studies. This adoption included only the adaptation of a single instance virtual machine for computational tasks.

3.2.4 Consideration of Shading Effects

Observed in approximately 84.21% of the 76 reviewed literatures, shading effects has been one of the key considerations. The consideration of shading effects was reflected either through specialised, estimation, or elimination techniques. Specialised shading methods included the use of algorithms, modules, software, and software extensions in the analysis. Shading effect consideration through estimation included utilisation of shading coefficients, correction formula, reduction factors. While considerations of shading effects through elimination includes determination of omission criteria or the removal of features that do not meet the requirement as it could potentially be shaded. However, the most prevalent approach in considering shading in solar energy planning was the utilisation of shading coefficient, which varies from 0.35 to 0.959.

3.2.5 Integration of Datasets into a Database

Integrating datasets into a database was only observed in 2.63% of the reviewed literatures. From the studies reviewed, the utilisation of database was observed to be either for the storage of information regarding PV potential information about each roof surface, or to conduct data augmentation and transfer learning strategies.

3.2.6 Utilisation of Solar Energy Calculation Models

Solar energy calculation models play an integral role in the evaluation of the feasibility of solar deployments, evident in around 81.58% of the reviewed literatures. Physical Mathematical Models were widely used to estimate the physical potential for energy generation. The solar energy potential model developed by Hoogwijk [39], as seen in Eq. (1), was extensively used in the reviewed literatures, with multiple variations of this formula being observed.

$$E_i = G_i \cdot \eta_m \cdot pr \quad (1)$$

where: E_i is the PV electricity production in grid cell i,
G_i is the geographical potential (kWh).
η_m is the conversion efficiency of the PV module.
pr is the performance ratio of the PV module.

3.2.7 Visualisation of Solar Energy Planning Results with Web-Mapping Interface

Approximately 7.89% of the reviewed papers utilised web-mapping interfaces to present their results in solar energy planning. The deployments varied from utilisation of OpenLayers with WMS file formats, WebGIS viewer to deployment of online mapping applications or platforms with a working graphical user interface (GUI).

3.2.8 Recommendations for Policy and Planning Strategies

In 57.89% of the reviewed papers, strategies involving policy, planning or a combination of both were identified to optimise resources and enhance urban solar rooftop energy generation. Recommendations were primarily geared towards presenting the planning strategy, such as optimal rooftop locations for deployment, tilt and orientation of PV arrays for maximum electricity production.

3.2.9 Inclusion of Environmental, Social, and Economic Factors

The integration of environmental, social, and economic factors was found to be present in approximately 32.89% of the reviewed literatures. Due to the varying motivations in energy generation through solar, different studies integrate a variety of factors based on their specific objective, with some focusing only one factor and the others on a selection of factors. Table 2 includes the list of collated factors considered in solar energy planning within the reviewed literatures, highlighting the economic, social and environmental considerations. Most of the factors in planning for solar energy production, aside from the solar radiation and electricity potential, are economic factors.

3.2.10 Tracking of Net-Zero Emission Goals

Applying CO_2 emission calculation models to calculate or project net-zero emission goals offset in solar energy planning is only present in 15% of the reviewed literatures. Numerous ways were observed to obtain emission offset values. Approaches include the use of software including RETScreen and SimaPro, utilisation of calculation models, applying CO_2 reduction coefficient, conducting LCA methodology with

ISO 14040/14044 framework, and employing the Carbon intensity (CI) conversion rate.

3.2.11 Applications of Digital Twin Technology

From the reviewed literatures, there were no observed application of digital twin technologies.

4 Discussions and Future Research Agenda

4.1 Quantitative Network Analysis Overview

4.1.1 Co-citation Networks

Based on the scientometric results depicted in Fig. 3b, Ordenes et al. [80] has the highest centrality despite being co-cited only 3 times, while Izquierdo et al. [46], with the most co-citations, holds the second highest centrality score. This discrepancy reflects the distinct roles these studies play within the network.

Ordenes et al. [80] investigated the impact of building integrated photovoltaics (BIPV) on the energy demand of Brazilian multi-family homes. In their study, a simulation of buildings in three different cities using the Simple Model of EnergyPlus software to forecast patterns of different PV systems under various temperatures. A prototype was developed that includes adapting weather and geography data, building model and PV system specifications. The results showed that even with the peak electricity demand of buildings and lowest solar radiation, the energy production from PV will still suffice. On the other hand, Izquierdo et al. [46] estimated solar energy potential within roofs for large-scale deployment using a hierarchical approach. Their research influenced the feasibility assessment of solar adoption by highlighting the necessity for an integrated and holistic approach.

Additionally, Ordenes et al. [80] served as a critical author, connecting the main theme "#0 rooftop PV" that facilitated the flow to the new theme "#4 experimental measure", which subsequently contributed to the prominence of now critical theme in the industry, "#2 renewable energy policies." While Izquierdo et al. [46] is more frequently cited within a concentrated thematic area especially in the "#0 rooftop pv" cluster.

Furthermore, extensions to themes "#10 non-ideal position", "#8 energy analysis", "#3 shading", "#5 sky view factor", "#7 lcoe of photovoltaics", "#1 urban morphology", "#18 agent-based modeling", "#11 photovoltaic" and "#18 stand-alone pv system" has been identified. These extensions provide deeper understanding for the main cluster regarding rooftop PV, contributing to its significance and complexity.

Additionally, key links connect the "#4 experimental measurements" cluster to the "renewable energy policies". This connection is observed to be due influence of the studies conducted by Nann and Emery [73] and Brinkworth and Sandberg [16].

Nann and Emery [73] had developed a model that calculates the hourly solar irradiance for both clear and cloudy skies from location-specific meteorological data. The researchers investigated power sensitivity and energy rating, which lead to the proposal of a multi-value energy rating scheme for critical operations times. Consecutively, Brinkworth and Sandberg [16], serving as the connecting authors to the "#2 renewable energy policies" cluster, modelled the optimal structural design of PV technology to minimise efficiency losses. The work conducted was a part of the European Union Fifth Collaboration PV Cool Build, highlighting the significant impact with the extended connection from "#4 experimental measurements". Their contribution has helped advance urban solar energy planning, particularly, in the development of renewable energy policies.

The analysis of the existing network reveals two significant gaps in the study. Firstly, there are little to no signs of connection between the "#2 renewable energy policies" cluster to the main cluster "#0 rooftop pv". This highlights a significant gap between the implementation of renewable energy policies and the planning of solar rooftop energy within urban and energy planning frameworks, potentially hindering the optimal integration of solar technologies into urban landscapes. This gap widens information silos within the industry and poses challenges in meeting environmental targets and formulating targeted policies for planning and deployment of renewable energy solutions.

Lastly, there are also little to no apparent links between "#4 experimental measurements" and the other extended themes of the main cluster #0, namely, "#8 energy analysis", "#5 sky view factor", "#1 urban morphology", "#18 agent-based modeling", "#11 photovoltaic" and "#18 stand-alone pv system". This highlights a currently limited and isolated test in conducting research that leads to suboptimal renewable energy policies as its current connections is only visible to the clusters "#3 shading", "#7 lcoe of photovoltaics" and "#10 non-ideal position". An explanation for this could be the varying background of the researchers and their domains of study in addressing challenges in urban energy planning. This fragmentation indicates a need to conduct experimental measurements that involve broader thematic areas to eliminate the limited scope and applicability of findings. As the limited scope could hinder the development of models and simulations that informs sustainable policies and deployments.

To bridge these gaps in solar energy planning, future studies could focus on collaborating and consider knowledge sharing to develop a holistic approach in tackling the challenges identified. By establishing holistic approaches, facilitation of a more comprehensive models that better inform decision-making process towards sustainable energy systems.

4.1.2 Co-occurrence Networks

Figure 4 highlights the progression of focus that shaped urban solar energy planning domain over time. Innovations in photovoltaic cell technologies and their implementation for a greener community to address climate issues were the starting points in the research landscape. It is reflected in the top 5 keywords that have the highest betweenness centrality alongside the keywords such as "alternative energy", "environmental impact", "carbon dioxide" and "emission control" between the years 1996 to 2004. Emphasising the global commitment to energy transition and reduction of reliance on fossil fuels. Then the focus has been observed to have shifted gradually towards the research of solar energy optimisation and planning, as reflected in keywords such as"gis", "building", "urban area", "cost effective", "cost benefit analysis", "assessment method", "sustainability" and "energy management". The shift emphasises the commitment to addressing societal concerns by integrating renewable energy solutions into sustainable development initiatives.

However, as cities evolve and technologies advance, the research landscape also undergoes transformation, presenting unique challenges. The emergence of words such as "exhibition", "energy modelling", "digital surface models", "information systems", "agent based modelling", "image segmentation", "multi objective optimization", "simulation platform" and "computation methods" starting from the early 2000 onwards indicated a shift towards sophisticated and computational approaches to urban solar energy assessments. Furthermore, this reflects a growing emphasis on leveraging advanced technologies and approaches in assessing and planning for urban energy.

It is evident that the dynamics of the nature of urban solar energy planning have shifted, highlighting the need for innovative strategies and tools to address rising challenges. Shifting towards computational methods through its integration in urban solar energy planning framework will ensure sustainability and efficiency in energy management in a rapidly changing urban environment.

4.2 Analytical Assessment Overview

4.2.1 Advantages and Challenges in 3D Solar Irradiation Modelling in Urban Environment

Studies reviewed that have employed 3D modelling for solar irradiation have presented significant advantages over traditional 2D approach. These includes: (1) the ability to capture precise solar values reflective of the real-world (Fig. 5a); (2) better consideration for the complexity of the urban environment that impacts solar insolation, such as differences in height (Fig. 5b) and impacts of shading; and (Fig. 5c) improved spatial resolution for more detailed solar energy potential (Fig. 5c).

One of the prominent approaches within the reviewed literatures regarding 3D solar radiation was the utilisation of LiDAR point cloud to reconstruct the area

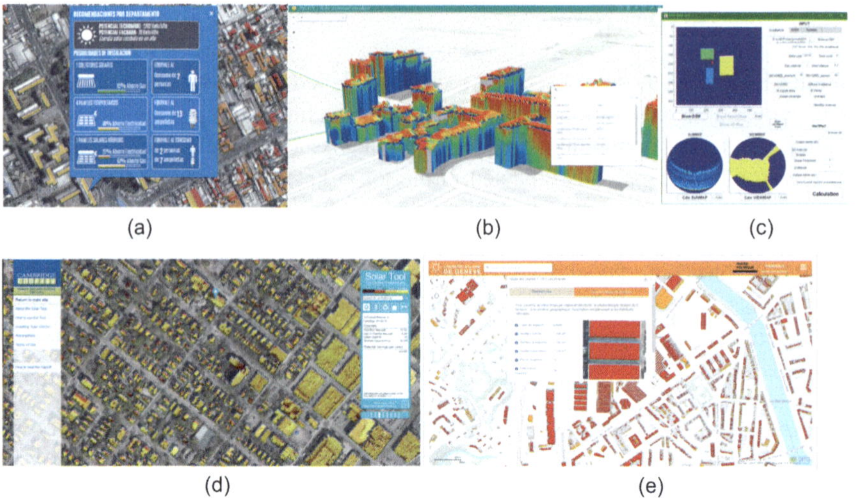

Fig. 5 Web-maps for urban rooftop PV **a** Web-map for the City of Conception, Chile [4], **b** Project WebGIS for Lisbon [17], **c** Graphical User interface of the tool developed by Oh and Park [79]; **d** Web Solar map of Cambridge, MA [49], and **e** Web-based developed for Geneva [26]

of interest. This is due to the enablement of 3D modelling with high precision in different levels of detail essential for solar potential analysis. Aside from using LiDAR, photogrammetry-based geometrical reconstructions provide another avenue for capturing 3D structure information. Unlike LiDAR however, it is limited in terms of considering urban surroundings [109]. Compared to 3D models generated from LiDAR data, those derived from geometrically reconstructed photographs have limited versatility but are more cost efficient and accurate for small scale studies. However, it has been proven that with both approaches combined, a more accurate urban representation and characterisation can be achieved, as investigated by Martín-Jiménez et al. [65].

Along with approaches are the software for 3D solar irradiation modelling. The dominantly used in the reviewed literatures have been 3D Rhinoceros with Grasshopper, Ladybug, and Honeybee plugins. This tool offered advanced capabilities that analyse solar potential and simulate different scenarios as shown by authors listed in Table 2.

By leveraging 3D environments, researchers can capture more precise and detailed solar radiation values within urban environments increasing the reliability in the decision-making process. Adopting these advancements enables a deeper and accurate solar energy potential within complex urban environments.

Despite the advantages, common challenges associated with 3D irradiation modelling includes data availability, computational complexity, and model validation. These vary depending on various aspects like scope of study area and temporal scale requirement, adding another layer of complexity. Addressing these challenges requires further research and development to improve the applicability of

3D modelling solar energy assessments. Future directions may include advancements in data acquisition technologies, development of efficient processing algorithms and standardisation of modelling techniques.

4.2.2 Opportunities for Real-Time Data Integration for Enhanced Solar Energy Planning

From the review literature, the utilisation of real-time datasets in the current research landscape has been very limited. Real-time data integration offers significant opportunities to improve solar planning and management. With up-to-date continuous data, an enhanced accuracy in solar energy generation calculations or predictions can be obtained.

Traditional solar energy planning data often relies on historical data and Typical Meteorological Year (TMY), which is not representative of the current conditions. Consequently, this creates a generalised solar analysis which creates either underestimation or overestimation. By incorporating real-time data, solar energy calculations can be more accurate for decision-making and planning which leads to sustainable energy yields and policies.

At present, this type of data is only integrated in energy consumption, yet opportunity poses for meteorological conditions and PV system performance. To fully capitalize this opportunity presented by real-time data, future research should not only ensure that current meteorological conditions and PV system performance are represented, but also develop sophisticated tools that enable interpretation and interoperability of real-time data efficiency.

4.2.3 Implications for Leveraging Cloud Computing for Solar Energy Planning

Whilst there is a limited application of Cloud Computing Services in the current reviewed urban solar energy planning landscape, the specific use of virtual machines for computational tasks highlights its most obvious potential. Desthieux et al. [26] and Buffat et al. [18] have both demonstrated an approach to which cloud-based solutions can enhance data processing in solar analysis. Adopting cloud services for data processing allows more timely information for decision-making, ensuring that solar deployments are optimised.

Aside from that, cloud computing also offers advantages such as remote access and collaboration, simulation and modelling capabilities, flexibility and scalability, and possible extension to other technologies. Through the cloud, access to essential solar analysis tools and datasets would be possible regardless of location if there is access to the internet. With the capability to connect through the cloud, it enables collaboration between researchers, planners and stakeholders that ensure coordinated and informed outcomes. Simulation and modelling would also be possible with cloud integration in the solar energy planning process. This would greatly enhance the capability to analyse potential, conduct experimental scenarios and test possibilities that would produce reliable evaluations and outcomes with ease.

Furthermore, as previously mentioned in Sect. 4.1.1, one of the key gaps is that there is currently limited and isolated testing in the conducted research. This is where cloud computing could be applied, as it allows both scalability and flexibility. With the privilege to scale up or down, researchers can manage projects addressing the requirements dynamically. Lastly, cloud services could relate to other emerging technologies such as Internet of Things (IoT) and Artificial Intelligence (AI), which are observed to be widely used in the solar energy planning studies, as they allow real-time data streaming and task automation, respectively. With cloud computing extended with AI and IoTs, the overall efficiency and reliability of solar energy analysis and insights could better inform planners.

However, with the opportunities for streaming, automation, flexibility, and simulation in rooftop PV planning comes with challenges. The challenges include the need for specialist skills to fully maximise cloud services and ensuring compliance with current regulations and policies regarding data privacy.

4.2.4 Shading Dynamics and Effects on Solar Rooftop Energy Assessments

As reflected in both quantitative and qualitative studies, shading significantly impacts solar energy generation, adding another layer of complexity in urban solar rooftop energy planning. While the most common approach to considering shading involves shading coefficients to reduce computational tasks in solar energy potential determination, it often leads to overestimation and underestimation in solar potential values. In urban energy planning, this could result in suboptimal decisions, leading to inefficient resource allocation and underutilised solar energy resource. Economically, it also impacts stakeholders and investors by reduced viability and return on investment.

To address these challenges, it is highly recommended to explore other shading techniques that offer great accuracy. The specialised shading methods used by other researchers in this review, or the quantification shading analysis modelling tools and software as elaborated by Jakica [48] and as included by Freitas et al. [30] in their review, offer promising avenues for accurate assessment of shadowing. Furthermore, acknowledging the limitations of the current shading analysis methods opens possibilities in technological innovations driven by collaboration and understanding among researchers, planners and decision-makers.

4.2.5 Data Integration and Database Implementation Strategies

Data integration to databases has been a critical step in other solar energy studies in general, however, when it comes to urban solar rooftop energy planning, there was an observed minimal integration. Özdemir et al. [81] used an integrated database for the PV potential information about each roof surface. while Yan et al. [105] used a relational database PostgreSQL for data augmentation.

The integrated database in the study of Özdemir et al. [81] highlights the utility of a database to manage data. By querying information, such as building and roof surface information, Özdemir et al. [81] displayed how database could facilitate seamless data retrieval for analysis or decision-making. On the other hand, Yan et al. [105] had used PostgreSQL to develop a model for the calculations of solar irradiation on 3D environment. This displayed the versatility of database management systems in modelling for solar energy research, demonstrating another application of databases for solar rooftop energy planning.

Furthermore, the applicability of database offers opportunities for enhancing data integration in rooftop PV studies. Centralising diverse data into a database allows for standardisation on data formats allowing interoperability and sharing. In addition, the availability of specialised platforms such as 3DCityDB offers enhancement for 3D urban data, applying this into urban solar rooftop planning enables researchers to innovate applications through visualisation techniques. Incorporating this can accelerate energy transitions by providing insights for informed decision-making processes.

4.2.6 Solar Energy Modelling Approaches

The results in Sect. 3.2.6 indicate that physical mathematical models have been the most dominant approach amongst the reviewed literatures; however, relying solely on this type of model may not be advantageous, as it often oversimply the complexities of solar energy potential. Factors such as solar irradiance, conversion coefficients, and efficiency factor can significantly impact energy estimates, potentially leading to misleading decision-making.

Similar to the effects of misestimations in shading effects, inaccuracies in solar energy modelling can have substantial effects on economic and planning initiatives. To minimise this issue, it is suggested for future researchers to utilise computational models that employ algorithms, tools, and software to analyse the potential of solar rooftop PVs in urban areas. This would provide a more nuanced analysis of the feasibility of solar deployments, minimising the risks of misinformed decisions.

Despite the current progress achieved in solar energy modelling, challenges still exist, such as limited access to tools and software, scalability, and validation. Addressing these challenges would require innovations aimed at advancing software that is geared towards open-source applications, and computation efficiency.

4.2.7 Enhancing Accessibility and Interpretation for Solar Energy Planning

Though a minimal number of studies presented their solar energy planning results through a web-mapping interface, this integration exemplified the importance of accessibility for both stakeholders, and decision-makers. The reviewed literatures indicated that using user interface or platform for the visualisation of complex urban solar information is strategic, as it increases awareness and greater engagement. Web-mapping employs various data and visualisation techniques to help users understand spatial patterns and relationships related to solar energy data. Different web-map presented in the literature, including interactive interfaces, has used either 2D/3D heat maps or 3D surface plots (as seen in Fig. 5). In addition to displaying solar values in a colour scale, multiple data such as satellite imagery, and statistical information were observed to be often added as background and supplementary information to enhance data comprehensibility.

Furthermore, web-mapping allows for the easy dissemination of updated information as soon as up-to-date information is available. Given the dynamic nature of solar resources, a flexible dissemination method will play a vital role in enabling timely insights for solar planning decision-making. Despite great efforts to enhance accessible information, these efforts face challenges in data accuracy, validation, and interoperability. This is due to the lack of standard in terms of building web-map and visualising information for solar rooftop energy planning. Addressing these challenges would be crucial to promoting sustainable solar energy development to the stakeholders. In addition, future efforts should leverage emerging technologies that enable data analytics, stakeholder engagement and data modelling to enhance accessibility. A continued effort in innovating delivering insights will be essential for advancing sustainable energy management and achieving environmental goals.

4.2.8 Proposed Strategies for Policy Implementation and Urban Planning

The most common advice presented in the literature review involved planning strategy recommendations alone; however, alongside from delivering solar values essential for solar energy planning, proposing strategies for policy implementation is equally essential. The implications of any solar rooftop planning study should align to the urban planning goals and create policy promoting renewable energy transition. By guiding planners with spatial insights and evidence, cities can optimise the placement and integration of solar technologies. Furthermore, aligning policies to be comprehensive can support the deployment of solar technologies in urban environments effectively.

As a common trend within the review literature, identification of rooftops that has the highest potential and advising to target those key areas then followed by adding incentive mechanism has been a common trend when it comes to the proposed strategies involving energy planning. As cities evolve, changes in financial incentives would play a key role in the pace of PV adoption. Frequent analysis will be necessary to ensure the viability of solar investments.

Additionally, monitoring and evaluation mechanisms are critical to assessing the effectiveness of policy interventions in solar urban planning. Future research should focus on integrating measurement of impact and recommendations on policy and plan adjustments to achieve the intended goal.

4.2.9 Interplay of Environmental, Social, and Economic Factors

While economic factors have traditionally taken priority in the solar energy planning designing process, however, we contend that both environmental and social factors are equally essential. Nonetheless, we recognise the importance of carefully weighing the trade-offs to maintain economic viability. To address potential conflicts, it is crucial to identify strategies for alternate solutions that need to be presented to maximise both resources and promote sustainability. Future researchers for solar rooftop energy planning should aim to explore innovative methods that balance economic, environmental, and social objections, thereby facilitating both sustainable and include energy transitions.

4.2.10 Path to Net-Zero Emissions Tracking

Even amid the global urgency to address environmental concerns, only a few of the reviewed literatures has integrated carbon emission model or any environmental model to calculate or track emission goal offset in the solar rooftop energy assessments. Given the significant variations of emission in different locations, understanding current levels of emission and ability offset carbon is essential for establishing a baseline and conducting critical assessment of how targets can be achieved. This is where monitoring and reporting initiatives play a key role in tracking progress towards net-zero emission goals. Furthermore, it requires a standardised reporting framework and key performance indicators to assess reductions and identify areas for improvement. Not only will this ensure stakeholders remain transparent about the environmental goals, but also remain informed and accountable for emission reduction efforts. To overcome these challenges, consistent efforts and investments from both the government and other stakeholders are essential to strengthen regulations for monitoring environmental goals.

Future directions in net-zero emission tacking will involve further advancements in data analysis research and adaptation of remote sensing technologies to obtain reliable data for reliable analysis.

4.2.11 Potential of Digital Twins for Urban Rooftop Solar Energy Planning Future

No direct application of digital twins has been found from the reviewed literatures; however, many studies aimed for more accurate and reliable assessment, which aligns with the overarching goals of digital twins. For instance, several studies (as presented in Table 2) have employed methodologies and approaches that exhibit strong similarities to the principles of digital twins.

Furthermore, considering the future directions outlined in the previous section three main shifts need to be realised for a sustainable energy management: Shift to realistic data representation, shift to integrated solar modelling, and shift to imperative solar energy planning.

Firstly, the shift to realistic data representation; this involves utilisation of semantic 3D city models and dynamic representation of urban processes. By adopting such an approach, planners can gain a more accurate and comprehensive understanding of the urban environment, facilitating precise solar modelling and analysis.

Secondly, the shift in integrated solar modelling. This includes the incorporation of computational methods and technologies namely, cloud computing services, shading analysis, integrated database, and solar energy modelling approaches. This enables effective and efficient spatiotemporal solar potential irradiation modelling and calculations, thereby minimising uncertainty and limited modelling for solar assessment.

Lastly, the shift to imperative solar energy planning. This emphasises the inclusion of spatial explicit information, dissemination tools for stakeholder engagement, presentation of policy and planning strategies, consideration of social, environmental and economic factors, and integration of carbon emission monitoring model for net-zero tracking.

Integrating digital twins to these can facilitate enhanced decision making. While the current application of digital twins in urban solar energy planning is currently limited, the potential for integration is promising, as evidenced by other fields of study.

4.3 Spatial Digital Twins Towards Sustainable Urban Rooftop Solar Energy Planning

Digital twinning as characterised by the replication of a physical object into the digital world, offers revolutionary potential and results for urban rooftop solar energy planning. Through this, time-consuming and human-intensive analysis tasks and actions can be conducted remotely. While there have been multiple studies that have investigated the potential of urban solar energy, there is no known study that specifically explores the real-time simulation of rooftop PV planning using interactive 3D environment, such as Leopold et al. [59] and Srinivasan et al. [92]. This poses a great opportunity with GIS and digital twins merged, for solar rooftop PV planning to advance towards sustainability.

Within the digital twin realm, spatial digital twin is one of the prominent digital innovations to be developed that can support realistic simulation, geographic analysis, and integrated decision-making in multiple scales. In the solar energy industry, this digital technology is beginning to manifest its potential.

Fig. 6 Aligned gemini principles to the future development of spatial digital twins for urban rooftop solar energy planning

Aligning with the GEMINI Principles of digital twins, as presented by Bolton et al. (2018), the future development of Spatial Digital Twins for Urban Rooftop Solar Energy Planning can represent several key principles, as illustrated in Fig. 6. By aligning these principles to future solar urban rooftop energy planning methodologies, enhanced data representation, model validation, and stakeholder engagement can be achieved.

5 Prospective Framework

Aimed at leveraging both the current initiatives presented in the literature, the existing challenges, and the principles mentioned in Sect. 4.3, a conceptual framework for urban rooftop planning is presented. This is illustrated in Fig. 7, which includes four key components, namely, Data, Analysis, Engagement, and Feedback.

Data forms the foundation of the spatial digital twin framework. For solar rooftop energy planning, necessary data includes 3D building data, meteorological data and sensor data. Analysis ensures various analytical processes are conducted to obtain insights. This involves solar modelling, shading effects, energy calculation and environmental, social, and economic implications. Engagement facilitates interaction between stakeholders and the virtual environment. It includes participatory planning and stakeholder engagement. Feedback informs the physical environment for insights and results to data being fed back to the system for a new cycle of reiteration.

With the presented framework, standardisation of methodologies allows both researchers and practitioners in solar rooftop PV planning to develop more efficient planning strategies. Accounting for these aspects in a framework brings significant benefits, such as identifying areas needing additional deployment and guiding the implementation of policies and planning strategies to optimise solar energy potential. Additionally, the proposed framework would serve as a guide towards realistic data representation, integrated approach in solar modelling and more imperative solar rooftop PV planning. Through the help of spatial digital twins, sustainable solar energy planning can be achieved.

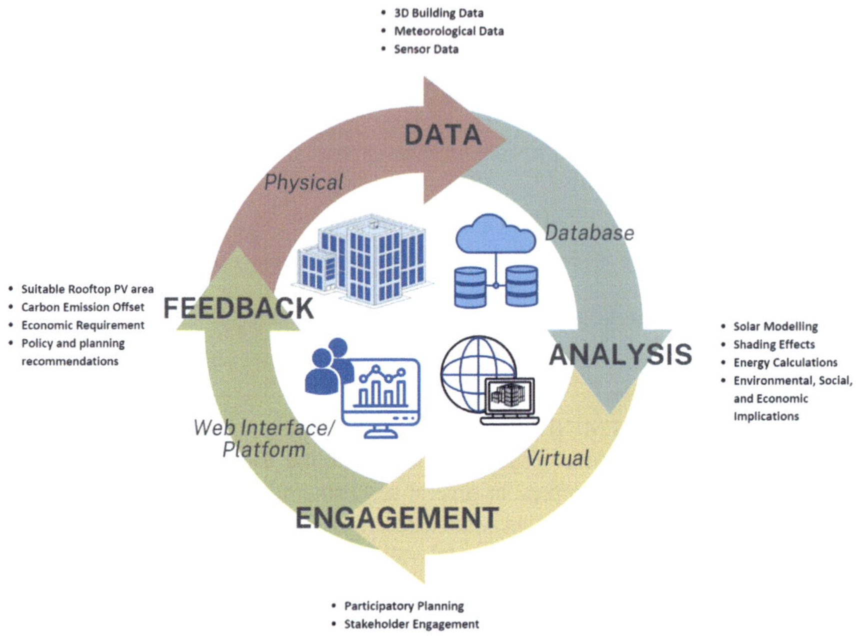

Fig. 7 Overview of the proposed high level conceptual framework for solar energy planning through spatial digital twins

6 Conclusion

This chapter highlights the areas of improvement in GIS-based methodologies for PV potential assessment and solar energy planning. The dual method allowed a comprehensive review of the current knowledge and approach to support effective sustainable solar rooftop adoption. The quantitative review revealed three key insights: a significant gap between renewable energy policies and existing rooftop PV deployment strategies, limitations in the scope of the current experimental measures for rooftop solar energy planning, and a trend towards more sophisticated computational approaches in urban rooftop solar energy planning.

The qualitative review highlighted three main shifts necessary for sustainable energy management through urban solar rooftop energy planning. Firstly, shift from a generalised urban representation to a more realistic data that incorporates semantic 3D city information models and dynamic representation of urban processes. Secondly, move from simplified urban modelling to more integrated modelling, utilising computational methods and technologies, such as cloud computing, shading analysis, integrated databases, and advanced solar energy modelling approaches. Lastly, the shift from abstracted spatial planning to an imperative spatial planning concept. This involves spatial explicit information, dissemination tools, policy and planning strategies recommendations, consideration of social, environmental and economic factors, and integration of carbon emission monitoring model for net-zero tracking.

These shifts will be possible through the adaptation of spatial digital twin technologies to provide a collaborative and sustainable solution for rooftop PV energy adoption on a city scale. Moreover, the future development of spatial digital twins for urban rooftop solar energy planning is viewed to align with Gemini principles 1, 2, 3, 5, 6, 7, and 9. The proposed conceptual framework can serve as a guide towards sustainable energy management, driving more effective, inclusive and comprehensive solar rooftop PV planning and implementations.

References

1. Adjiski V, Kaplan G, Mijalkovski S (2023) Assessment of the solar energy potential of rooftops using LiDAR datasets and GIS based approach [Article]. Int J Eng Geosci 8(2):188–199. https://doi.org/10.26833/ijeg.1112274
2. Akrofi MM, Okitasari M (2023) Beyond costs: How urban form could limit the uptake of residential solar PV systems in low-income neighborhoods in Ghana [Article]. Energy Sustain Dev 74:20–33. https://doi.org/10.1016/j.esd.2023.03.004
3. Al-Shahri OA, Ismail FB, Hannan MA, Lipu MSH, Al-Shetwi AQ, Begum RA, Al-Muhsen NFO, Soujeri E (2021) Solar photovoltaic energy optimization methods, challenges and issues: a comprehensive review. J Clean Prod 284. https://doi.org/10.1016/j.jclepro.2020.125465
4. Alvarado RG, Troncoso L, Campos P (2016) Residential solar energy potential for public dissemination: a case study in concepcion, Chile [Article]. J Green Build 11(1):118-U114. https://doi.org/10.3992/jgb.11.1.118.1
5. An YN, Chen TY, Shi L, Heng CK, Fan JL (2023) Solar energy potential using GIS-based urban residential environmental data: a case study of Shenzhen, China [Article]. Sustain Cities Soc 93:20. Article 104547. https://doi.org/10.1016/j.scs.2023.104547
6. Asif M, Hassanain MA, Nahiduzzaman KM, Sawalha H (2019) Techno-economic assessment of application of solar PV in building sector: a case study from Saudi Arabia [Article]. Smart Sustain Built Environ 8(1):34–52. https://doi.org/10.1108/Sasbe-06-2018-0032
7. Assouline D, Mohajeri N, Scartezzini JL (2017) Quantifying rooftop photovoltaic solar energy potential: a machine learning approach [Article]. Sol Energy 141:278–296. https://doi.org/10.1016/j.solener.2016.11.045
8. Assouline D, Mohajeri N, Scartezzini JL (2018) Large-scale rooftop solar photovoltaic technical potential estimation using Random Forests [Article]. Appl Energy 217:189–211. https://doi.org/10.1016/j.apenergy.2018.02.118
9. Ayodele TR, Ogunjuyigbe ASO, Nwakanma KC (2021) Solar energy harvesting on building's rooftops: a case of a Nigeria cosmopolitan city [Article]. Renew Energy Focus 38:57–70. https://doi.org/10.1016/j.ref.2021.06.001
10. Baghani A (2023) Assessment of rooftop solar power potential in rural areas using uav photogrammetry and gis [Article]. Renew Energy Res Appl 4(2):251–258. https://doi.org/10.22044/rera.2023.12555.1192
11. Bazan J, Rieradevall J, Gabarrell X, Vazquez-Rowe I (2018) Low-carbon electricity production through the implementation of photovoltaic panels in rooftops in urban environments: a case study for three cities in Peru [Article]. Sci Total Environ 622–623:1448–1462. https://doi.org/10.1016/j.scitotenv.2017.12.003
12. Bergamasco L, Asinari P (2011) Scalable methodology for the photovoltaic solar energy potential assessment based on available roof surface area: further improvements by Ortho-image analysis and application to Turin (Italy) [Article]. Sol Energy 85(11):2741–2756. https://doi.org/10.1016/j.solener.2011.08.010

13. Bernadette D, Twizerimana M, Bakundukize A, Jean Pierre B, Theoneste N (2021) Analysis of shading effects in solar pv system. Int J Sustain Green Energy 10(2) https://doi.org/10.11648/j.ijrse.20211002.13
14. Borfecchia F, Caiaffa E, Pollino M, De Cecco L, Martini S, La Porta L, Marucci A (2014) Remote sensing and GIS in planning photovoltaic potential of urban areas [Article]. Eur J Remote Sens 47:195–216. https://doi.org/10.5721/EuJRS20144713
15. Boulahia M, Djiar KA, Amado M (2021) Combined engineering—statistical method for assessing solar photovoltaic potential on residential rooftops: case of Laghouat in Central Southern Algeria. Energies 14(6) (undefined). https://doi.org/10.3390/en14061626
16. Brinkworth BJ, Sandberg M (2006) Design procedure for cooling ducts to minimise efficiency loss due to temperature rise in PV arrays. Solar Energy 80(1):89–103. https://doi.org/10.1016/j.solener.2005.05.020
17. Brito MC, Redweik P, Catita C, Freitas S, Santos M (2019) 3D Solar potential in the urban environment: a case study in Lisbon [Article]. Energies 12(18):13. Article 3457. https://doi.org/10.3390/en12183457
18. Buffat R, Grassi S, Raubal M (2018) A scalable method for estimating rooftop solar irradiation potential over large regions [Article]. Appl Energy 216:389–401. https://doi.org/10.1016/j.apenergy.2018.02.008
19. Chen BY, Che YB, Wang JK, Li HF, Yu LJ, Wang DC (2022) An estimation framework of regional rooftop photovoltaic potential based on satellite remote sensing images [Article]. Glob Energy Interconnect China 5(3):281–292. https://doi.org/10.1016/j.gloei.2022.06.006
20. Chen C (2005) CiteSpace II: Detecting and visualizing emerging trends and transient patterns in scientific literature. J Am Soc Inform Sci Technol 57(3):359–377. https://doi.org/10.1002/asi.20317
21. Chow A, Li S, Fung AS (2016) Modeling urban solar energy with high spatiotemporal resolution: a case study in Toronto, Canada [Article]. Int J Green Energy 13(11):1090–1101. https://doi.org/10.1080/15435075.2016.1170686
22. Cieślak I, Eźlakowski B (2023) Use of gis tools for decision-making support in sustainable energy generation on the example of the solar photovoltaic technology [Article]. Bulletin of Geography. Socio-Econ Ser (60). 157–171. https://doi.org/10.12775/bgss-2023-0021
23. Cuesta-Fernández I, Vargas-Salgado C, Alfonso-Solar D, Gómez-Navarro T (2023) The contribution of metropolitan areas to decarbonize the residential stock in Mediterranean cities: a GIS-based assessment of rooftop PV potential in Valencia, Spain [Article]. Sustain Cities Soc 97:12. Article 104727. https://doi.org/10.1016/j.scs.2023.104727
24. Davybida L, Wyczałek I, Plichta A (2022) Using GIS and SDSS tools in the design of a photovoltaic system for a built-up roof [Article]. GeomatS Environ Eng 16(4):31–58. https://doi.org/10.7494/geom.2022.16.4.31
25. Dehwah AHA, Asif M, Rahman MT (2018) Prospects of PV application in unregulated building rooftops in developing countries: A perspective from Saudi Arabia [Article]. Energy Build 171:76–87. https://doi.org/10.1016/j.enbuild.2018.04.001
26. Desthieux G, Carneiro C, Camponovo R, Ineichen P, Morello E, Boulmier A, Abdennadher N, Dervey S, Ellert C (2018) Solar energy potential assessment on rooftops and facades in large built environments based on lidar data, image processing, and cloud computing. methodological background, application, and validation in geneva (Solar Cadaster) [Article]. Front Built Environ 4:22. Article 14. https://doi.org/10.3389/fbuil.2018.00014
27. Doorga JRS, Tannoo R, Rughooputh SDDV, Boojhawon R (2021) Exploiting the rooftop solar photovoltaic potential of a tropical island state: case of the Mascarene Island of Mauritius [Article]. Int J Energy Environ Eng 12(3):401–418. https://doi.org/10.1007/s40095-020-00375-9
28. Fakhraian E, Alier M, Dalmau FV, Nameni A, Guerrero JC (2021) The urban rooftop photovoltaic potential determination [Article]. Sustainability (Switzerland) 13(13). Article 7447. https://doi.org/10.3390/su13137447
29. Fogl M, Moudry V (2016) Influence of vegetation canopies on solar potential in urban environments [Article]. Appl Geogr 66:73–80. https://doi.org/10.1016/j.apgeog.2015.11.011

30. Freitas S, Catita C, Redweik P, Brito MC (2015) Modelling solar potential in the urban environment: State-of-the-art review. Renew Sustain Energy Rev 41:915–931. https://doi.org/10.1016/j.rser.2014.08.060
31. Fuster-Palop E, Prades-Gil C, Masip X, Viana-Fons J D, Paya J (2023) Techno-economic potential of urban photovoltaics: comparison of net billing and net metering in a mediterranean municipality [Article]. Energies 16(8):32. Article 3564. https://doi.org/10.3390/en16083564
32. Gagnon P, Margolis R, Melius J, Phillips C, Elmore R (2018) Estimating rooftop solar technical potential across the US using a combination of GIS-based methods, lidar data, and statistical modeling [Article]. Environ Res Lett 13(2). Article 024027. https://doi.org/10.1088/1748-9326/aaa554
33. Garg B, Barach R (2021) Collaborative governance for urban sustainability: implementing solar cities [Article]. Asia Pac J Public Adm 43(4):236–257. Article 1925132. https://doi.org/10.1080/23276665.2021.1925132
34. Gawley D, McKenzie P (2022) Investigating the suitability of GIS and remotely-sensed datasets for photovoltaic modelling on building rooftops [Article]. Energy Build 265:14. Article 112083. https://doi.org/10.1016/j.enbuild.2022.112083
35. Gharakhani Siraki A, Pillay P (2012) Study of optimum tilt angles for solar panels in different latitudes for urban applications [Article]. Sol Energy 86(6):1920–1928. https://doi.org/10.1016/j.solener.2012.02.030
36. Gooding J, Edwards H, Giesekam J, Crook R (2013) Solar City Indicator: a methodology to predict city level PV installed capacity by combining physical capacity and socio-economic factors [Article]. Sol Energy 95:325–335. https://doi.org/10.1016/j.solener.2013.06.027
37. Han JY, Chen YC, Li SY (2022) Utilising high-fidelity 3D building model for analysing the rooftop solar photovoltaic potential in urban areas [Article]. Sol Energy 235:187–199. https://doi.org/10.1016/j.solener.2022.02.041
38. Hong T, Lee M, Koo C, Jeong K, Kim J (2017) Development of a method for estimating the rooftop solar photovoltaic (PV) potential by analyzing the available rooftop area using Hillshade analysis [Article]. Appl Energy 194:320–332. https://doi.org/10.1016/j.apenergy.2016.07.001
39. Hoogwijk MM (2004) On the global and regional potential of renewable energy sources. Universiteit Utrecht. Utrecht
40. Huang Y, Chen ZQ, Wu B, Chen L, Mao WQ, Zhao F, Wu JP, Wu JH, Yu BL (2015) Estimating roof solar energy potential in the downtown area using a gpu-accelerated solar radiation model and airborne lidar data [Article]. Remote Sens 7(12):17212–17233. https://doi.org/10.3390/rs71215877
41. Hubinsky T, Hajtmanek R, Seligová A, Legény J, Spacek R (2023) Potentials and limits of photovoltaic systems integration in historic urban structures: the case study of monument reserve in bratislava, slovakia [Article]. Sustainability 15(3):27. Article 2299. https://doi.org/10.3390/su15032299
42. International Energy Agency, IEA (2008) World Energy Outlook 2008. Retrieved February 7, 2024, from https://www.iea.org/reports/world-energy-outlook-2008.
43. International Energy Agency, IEA (2023) World Energy Outlook 2023. Retrieved February 7, 2024, from https://www.iea.org/reports/world-energy-outlook-2023
44. International Energy Agency, IEA (2024) Electricity 2024 Analysis and forecast to 2026. Retrieved February 7, 2024, from https://iea.blob.core.windows.net/assets/6b2fd954-2017-408e-bf08-952fdd62118a/Electricity2024-Analysisandforecastto2026.pdf
45. International Renewable Energy Agency, IRENA (2020) Rise of renewables in cities: energy solutions for the urban future. International Renewable Energy Agency, Abu Dhabi. Retrieved February 26, 2024, from https://www.irena.org/-/media/Files/IRENA/Agency/Publication/2020/Oct/IRENA_Renewables_in_cities_2020.pdf
46. Izquierdo S, Rodrigues M, Fueyo N (2008) A method for estimating the geographical distribution of the available roof surface area for large-scale photovoltaic energy-potential evaluations [Article]. Sol Energy 82(10):929–939. https://doi.org/10.1016/j.solener.2008.03.007

47. Jaglin S (2023) Urban electric hybridization: exploring the politics of a just transition in the western cape (South Africa) [Article]. J Urban Technol 30(2):11–33. https://doi.org/10.1080/10630732.2022.2111176
48. Jakica N (2018) State-of-the-art review of solar design tools and methods for assessing daylighting and solar potential for building-integrated photovoltaics. Renew Sustain Energy Rev 81:1296–1328. https://doi.org/10.1016/j.rser.2017.05.080
49. Jakubiec JA, Reinhart CF (2013) A method for predicting city-wide electricity gains from photovoltaic panels based on LiDAR and GIS data combined with hourly Daysim simulations [Article]. Sol Energy 93:127–143. https://doi.org/10.1016/j.solener.2013.03.022
50. Jo JH, Rose Z, Cross J, Daebel E, Verderber A, Kostelnick JC (2015) Application of airborne lidar data and geographic information systems (gis) to develop a distributed generation system for the town of normal, IL [Article]. Aims Energy 3(2):173–183. https://doi.org/10.3934/energy.2015.2.173
51. Kaleshwarwar A, Bahadure S (2023) Assessment of the solar energy potential of diverse urban built forms in Nagpur, India [Article]. Sustain Cities Soc 96:15. Article 104681. https://doi.org/10.1016/j.scs.2023.104681
52. Karteris M, Slini T, Papadopoulos AM (2013) Urban solar energy potential in Greece: a statistical calculation model of suitable built roof areas for photovoltaics [Article]. Energy Build 62:459–468. https://doi.org/10.1016/j.enbuild.2013.03.033
53. Khan J, Arsalan MH (2016) Estimation of rooftop solar photovoltaic potential using geo-spatial techniques: a perspective from planned neighborhood of Karachi—Pakistan [Article]. Renew Energy 90:188–203. https://doi.org/10.1016/j.renene.2015.12.058
54. Khan J, Arsalan MH (2016) Solar power technologies for sustainable electricity generation—a review. Renew Sustain Energy Rev 55:414–425. https://doi.org/10.1016/j.rser.2015.10.135
55. Khan M, Asif M, Stach E (2017) Rooftop PV potential in the residential sector of the kingdom of saudi arabia. Buildings 7(4) (undefined). https://doi.org/10.3390/buildings7020046
56. Kouhestani FM, Byrne J, Johnson D, Spencer L, Hazendonk P, Brown B (2019) Evaluating solar energy technical and economic potential on rooftops in an urban setting: the city of Lethbridge, Canada [Article]. Int J Energy Environ Eng 10(1):13–32. https://doi.org/10.1007/s40095-018-0289-1
57. Kozlovas P, Gudzius S, Ciurlionis J, Jonaitis A, Konstantinaviciute I, Bobinaite V (2023) Assessment of technical and economic potential of urban rooftop solar photovoltaic systems in Lithuania [Article]. Energies 16(14):29. Article 5410. https://doi.org/10.3390/en16145410
58. Kumar P, Kumar P, Aggarwal V (2023) A behavioral study on villagers' adoption intention and carbon neutrality toward rooftop solar photovoltaic systems in India [Article; Early Access]. Int J Energy Sector Manage 28. https://doi.org/10.1108/Ijesm-11-2021-0009
59. Leopold U, Braun C, Pinheiro P (2023) An interoperable digital twin to simulate spatio-temporal photovoltaic power output and grid congestion at neighbourhood and city levels in Luxembourg. 2023 Free and open source software for geospatial, FOSS4G 2023
60. Liu X, Wang X, Wright G, Cheng J, Li X, Liu R (2017) A state-of-the-art review on the integration of building information modeling (bim) and geographic information system (GIS). ISPRS Int J Geo-Inf 6(2). https://doi.org/10.3390/ijgi6020053
61. Lukac N, Seme S, Zlaus D, Stumberger G, Zalik B (2014) Buildings roofs photovoltaic potential assessment based on LiDAR (Light Detection And Ranging) data [Article]. Energy 66:598–609. https://doi.org/10.1016/j.energy.2013.12.066
62. Machete R, Falcao AP, Gomes MG, Rodrigues AM (2018) The use of 3D GIS to analyse the influence of urban context on buildings' solar energy potential [Article]. Energy Build 177:290–302. https://doi.org/10.1016/j.enbuild.2018.07.064
63. Mangiante MJ, Whung PY, Zhou L, Porter R, Cepada A, Campirano E, Licon D, Lawrence R, Torres M (2020) Economic and technical assessment of rooftop solar photovoltaic potential in Brownsville, Texas, U.S.A [Article]. Comput Environ Urban Syst 80:1–101450. Article 101450. https://doi.org/10.1016/j.compenvurbsys.2019.101450
64. Margolis R, Gagnon P, Melius J, Phillips C, Elmore R (2017) Using GIS-based methods and lidar data to estimate rooftop solar technical potential in US cities. Environ Res Lett 12(7) (undefined). https://doi.org/10.1088/1748-9326/aa7225

65. Martín-Jiménez J, Del Pozo S, Sánchez-Aparicio M, Lagüela S (2020) Multi-scale roof characterization from LiDAR data and aerial orthoimagery: Automatic computation of building photovoltaic capacity [Article]. Autom Constr 109:14. Article 102965. https://doi.org/10.1016/j.autcon.2019.102965
66. Massano M, Macii E, Lanzini A, Patti E, Bottaccioli L (2023) A GIS open-data co-simulation platform for photovoltaic integration in residential urban areas [Article]. Engineering 26:198–213. https://doi.org/10.1016/j.eng.2022.06.020
67. McIntyre JH (2012) Community-scale assessment of rooftop-mounted solar energy potential with meteorological, atlas, and GIS data: a case study of Guelph, Ontario (Canada) [Article]. Energy Sustain Soc 2:19. Article 23. https://doi.org/10.1186/2192-0567-2-23
68. Melius J, Margolis R, Ong S (2013) Estimating rooftop suitability for PV: a review of methods, patents, and validation techniques. National Renewable Energy Laboratory Prepared under Task No. SS13.1010
69. Mikovits C, Schauppenlehner T, Scherhaufer P, Schmidt J, Schmalzl L, Dworzak V, Hampl N, Sposato RG (2021). A spatially highly resolved ground mounted and rooftop potential analysis for photovoltaics in Austria [Article]. Isprs Int J Geo-Inf 10(6):15. Article 418. https://doi.org/10.3390/ijgi10060418
70. Mishra T, Rabha A, Kumar U, Arunachalam K, Sridhar V (2020) Assessment of solar power potential in a hill state of India using remote sensing and geographic information system [Article]. Remote Sens Appl-Soc Environ 19:10. Article 100370. https://doi.org/10.1016/j.rsase.2020.100370
71. Moher D, Liberati A, Tetzlaff J, Altman DG, Group P (2009) Preferred reporting items for systematic reviews and meta-analyses: the PRISMA statement. BMJ 339:b2535. https://doi.org/10.1136/bmj.b2535
72. Muhammed E, El-Shazly A, Morsy S (2023) Building rooftop extraction using machine learning algorithms for solar photovoltaic potential estimation [Article]. Sustainability 15(14):17. Article 11004. https://doi.org/10.3390/su151411004
73. Nann S, Emery K (1992) Spectral effects on PV-device rating. Solar Energy Mater Solar Cells 27(3):189–216. https://doi.org/10.1016/0927-0248(92)90083-2
74. Neves D, Baptista P, Rosa P (2023) Solar energy communities: an open data and georeferenced based modelling framework to pre-assess deployment potential at urban level [Article]. Energy 282. Article 128838. https://doi.org/10.1016/j.energy.2023.128838
75. Nguyen HT, Pearce JM (2012) Incorporating shading losses in solar photovoltaic potential assessment at the municipal scale [Article]. Sol Energy 86(5):1245–1260. https://doi.org/10.1016/j.solener.2012.01.017
76. Nguyen HT, Pearce JM (2013) Automated quantification of solar photovoltaic potential in cities [Article]. Int Rev Spat Plan Sustain Dev 1(1),:49–60. https://doi.org/10.14246/irspsd.1.1_49
77. Nguyen HT, Pearce JM, Harrap R, Barber G (2012) The application of LiDAR to assessment of rooftop solar photovoltaic deployment potential in a municipal district unit [Article]. Sensors (Basel) 12(4):4534–4558. https://doi.org/10.3390/s120404534
78. Ninsawat S, Hossain M (2016) Identifying potential area and financial prospects of rooftop solar photovoltaics (PV). Sustainability 8(10) (undefined). https://doi.org/10.3390/su8101068
79. Oh M, Park H.-D (2019) Optimization of solar panel orientation considering temporal volatility and scenario-based photovoltaic potential: a case study in seoul national university [Article]. Energies 12(17):17. Article 3262. https://doi.org/10.3390/en12173262
80. Ordenes M, Marinoski DL, Braun P, Rüther R (2007) The impact of building-integrated photovoltaics on the energy demand of multi-family dwellings in Brazil. Energy Build 39(6):629–642. https://doi.org/10.1016/j.enbuild.2006.10.006
81. Özdemir S, Yavuzdoğan A, Bilgilioğlu B B, Akbulut Z (2023) SPAN: an open-source plugin for photovoltaic potential estimation of individual roof segments using point cloud data [Article]. Renew Energy 216:13. 119022. https://doi.org/10.1016/j.renene.2023.119022
82. Palmer D, Koumpli E, Cole I, Gottschalg R, Betts T (2018) A GIS-based method for identification of wide area rooftop suitability for minimum size PV systems using LiDAR data and

photogrammetry [Article]. Energies 11(12):22. Article 3506. https://doi.org/10.3390/en1112 3506
83. Petrasova A, Harmon B, Petras V, Tabrizian P, Mitasova H (2018) Viewshed analysis. In: Petrasova A, Harmon B, Petras V, Tabrizian P, Mitasova H (eds) Tangible modeling with open source GIS. Springer International Publishing, pp 127–132. https://doi.org/10.1007/978-3-319-89303-7_9
84. Quiquerez L, Faessler J, Lachal B, Mermoud F, Hollmuller P (2015) GIS methodology and case study regarding assessment of the solar potential at territorial level: PV or thermal? [Article]. Int J Sustain Energy Plan Manag 6:3–16. https://doi.org/10.5278/ijsepm.2015.6.2
85. Rachchh R, Kumar M, Tripathi B (2016) Solar photovoltaic system design optimization by shading analysis to maximize energy generation from limited urban area [Article]. Energy Convers Manag 115:244–252. https://doi.org/10.1016/j.enconman.2016.02.059
86. Ramirez Camargo L, Zink R, Dorner W, Stoeglehner G (2015) Spatio-temporal modeling of roof-top photovoltaic panels for improved technical potential assessment and electricity peak load offsetting at the municipal scale [Article]. Comput Environ Urban Syst 52:58–69. https://doi.org/10.1016/j.compenvurbsys.2015.03.002
87. Revesz M, Zamini S, Oswald SM, Trimmel H, Weihs P (2020) SEBEpv—new digital surface model based method for estimating the ground reflected irradiance in an urban environment [Article]. Sol Energy 199:400–410. https://doi.org/10.1016/j.solener.2020.01.075
88. Roth M, Tilk C (2016) Here comes the sun: a prototypical gis-based 4-dimensional rooftop solar potential analysis to support the energy turnaround [Article]. J Digit Landsc Arch 2016(1):133–140. https://doi.org/10.14627/537612016
89. Salim DHC, Mello CCD, Franco GG, Nóbrega RAD, de Paula EC, Fonseca BM, Nero MA (2023) Unveiling Fernando de Noronha Island's photovoltaic potential with unmanned aerial survey and irradiation modeling [Article]. Appl Energy 337:11. Article 120857. https://doi.org/10.1016/j.apenergy.2023.120857
90. Santos T, Gomes N, Freire S, Brito MC, Santos L, Tenedório JA (2014) Applications of solar mapping in the urban environment [Article]. Appl Geogr 51:48–57. https://doi.org/10.1016/j.apgeog.2014.03.008
91. Singh R, Banerjee R (2015) Estimation of rooftop solar photovoltaic potential of a city [Article]. Sol Energy 115:589–602. https://doi.org/10.1016/j.solener.2015.03.016
92. Srinivasan RS, Manohar B, Issa RRA (2020) Urban building energy CPS (UBE-CPS): real-time demand response using digital twin. In: Cyber-physical systems in the built environment. Springer International Publishing, pp 309–322. https://doi.org/10.1007/978-3-030-41560-0_17
93. Stack V, Narine L L (2022) Sustainability at Auburn University: assessing rooftop solar energy potential for electricity generation with remote sensing and gis in a southern us campus [Article]. Sustainability 14(2):14. Article 626. https://doi.org/10.3390/su14020626
94. Sultani RM, Soliman AM, Al-Hagla KS (2009) The use of geographic information system (gis) based spatial decision support system (sdss) in developing the urban planning process. Arch & Plan J 97–115. https://www.researchgate.net/profile/Khalid-Al-Hagla/publication/273457666_The_Use_of_Geographic_Information_System_GIS_Based_Spatial_Decision_Support_System_SDSS_in_Developing_the_Urban_Planning_Process/links/55031de50cf24cee39fd5a0d/The-Use-of-Geographic-Information-System-GIS-Based-Spatial-Decision-Support-System-SDSS-in-Developing-the-Urban-Planning-Process.pdf
95. Sun YW, Hof A, Wang R, Liu J, Lin YJ, Yang DW (2013) GIS-based approach for potential analysis of solar PV generation at the regional scale: a case study of Fujian Province. Energy Policy 58:248–259. https://doi.org/10.1016/j.enpol.2013.03.002
96. Thebault M, Clivillé V, Berrah L, Desthieux G (2020) Multicriteria roof sorting for the integration of photovoltaic systems in urban environments [Article]. Sustain Cities Soc 60:12. Article 102259. https://doi.org/10.1016/j.scs.2020.102259
97. Thebault M, Desthieux G, Castello R, Berrah L (2022) Large-scale evaluation of the suitability of buildings for photovoltaic integration: Case study in greater geneva [Article]. Appl Energy 316. Article 119127. https://doi.org/10.1016/j.apenergy.2022.119127

98. United Nations, D o E a S A, Population Division (2019) World urbanization prospects: the 2018 revision. https://population.un.org/wup/Publications/Files/WUP2018-Report.pdf
99. Uyan M (2013) GIS-based solar farms site selection using analytic hierarchy process (AHP) in Karapinar region, Konya/Turkey. Renew Sustain Energy Rev 28:11–17. https://doi.org/10.1016/j.rser.2013.07.042
100. Verso A, Martin A, Amador J, Dominguez J (2015) GIS-based method to evaluate the photovoltaic potential in the urban environments: The particular case of Miraflores de la Sierra [Article]. Sol Energy 117:236–245. https://doi.org/10.1016/j.solener.2015.04.018
101. Wang P, Yu P, Huang L, Zhang Y (2022) An integrated technical, economic, and environmental framework for evaluating the rooftop photovoltaic potential of old residential buildings. J Environ Manage 317:115296. Article 115296. https://doi.org/10.1016/j.jenvman.2022.115296
102. Widodo DA, Purwanto P, Hermawan H (2020) Potential of solar energy in residential rooftop surface area in semarang city, indonesia [Article]. Adv Sci, Technol Eng Syst J 5(4):397–404. https://doi.org/10.25046/aj050446
103. Wiginton LK, Nguyen HT, Pearce JM (2010) Quantifying rooftop solar photovoltaic potential for regional renewable energy policy. Comput Environ Urban Syst 34(4):345–357. https://doi.org/10.1016/j.compenvurbsys.2010.01.001
104. Wong MS, Zhu R, Liu Z, Lu L, Peng J, Tang Z, Lo CH, Chan WK (2016) Estimation of Hong Kong's solar energy potential using GIS and remote sensing technologies. Renew Energy 99:325–335. https://doi.org/10.1016/j.renene.2016.07.003
105. Yan LX, Zhu R, Kwan MP, Luo W, Wang D, Zhang SW, Wong MS, You LL, Yang BS, Chen BY, Feng L (2023) Estimation of urban-scale photovoltaic potential: a deep learning-based approach for constructing three-dimensional building models from optical remote sensing imagery imagery [Article]. Sustain Cities Soc 93:14. Article 104515. https://doi.org/10.1016/j.scs.2023.104515
106. Yang Y, Campana PE, Stridh B, Yan J (2020) Potential analysis of roof-mounted solar photovoltaics in Sweden [Article]. Appl Energy 279:14. Article 115786. https://doi.org/10.1016/j.apenergy.2020.115786
107. Yesilmaden HM, Dogru AO (2019) Finding the best locations for photovoltaic panel installation in urbanized areas [Article; Proceedings Paper]. Fresen Environ Bull 28(2):619–625. <Go to ISI>://WOS:000461270700064
108. Zalamea-Leon E, Mena-Campos J, Barragan-Escandon A, Parra-Gonzalez D, Mendez-Santos P (2018) Urban photovoltaic potential of inclined roofing for buildings in heritage centers in equatorial areas [Article]. J Green Build 13(3):45–69. https://doi.org/10.3992/1943-4618.13.3.45
109. Zhang W, Wong NH, Zhang Y, Chen Y, Tong S, Zheng Z, Chen J (2019) Evaluation of the photovoltaic potential in built environment using spatial data captured by unmanned aerial vehicles. Energy Sci Eng 7(5):2011–2025. https://doi.org/10.1002/ese3.408
110. Zhang Y, Schlueter A, Waibel C (2023) SolarGAN: synthetic annual solar irradiance time series on urban building facades via Deep Generative Networks [Article]. Energy AI 12. Article 100223. https://doi.org/10.1016/j.egyai.2022.100223
111. Zhong T, Zhang Z, Chen M, Zhang K, Zhou Z, Zhu R, Wang Y, Lü G, Yan J (2021) A city-scale estimation of rooftop solar photovoltaic potential based on deep learning. Appl Energy 2980. Article 117132. https://doi.org/10.1016/j.apenergy.2021.117132
112. Zhu R, Kwan M-P, Perera ATD, Fan H, Yang B, Chen B, Chen M, Qian Z, Zhang H, Zhang X, Yang J, Santi P, Ratti C, Li W, Yan J (2023) GIScience can facilitate the development of solar cities for energy transition. Adv Appl Energy 10. https://doi.org/10.1016/j.adapen.2023.100129
113. Zubair M, Ghuffar S, Shoaib M, Awan A B, Bhatti A R (2020) Assessment of photovoltaic capabilities in urban environments: a case study of Islamabad, Pakistan [Article]. J Sol Energy Eng-Trans Asme 142(6):13. Article 061006. https://doi.org/10.1115/1.4046947

Creating a Foundation for Spatial Digital Twins in Data-Scarce Regions Through Open-Source Solutions

Nicholas Lee, Chayn Sun, Monica Wachowicz, Debaditya Acharya, and Serene Ho

Abstract This chapter explored the extent that open-source solutions can help generate a spatial Digital Twin of cities in data-scarce regions—in this instance, Honiara, the capital city of Solomon Islands. City-model spatial Digital Twins are defined as a virtual twin of an existing city that has a channel of live data between the virtualised environment and physical world. The conceptual minimum requirements were investigated and whether the available data could define the city-model as a spatial Digital Twin. A city-model was generated with an Application Programming Interface (API) installed to communicate updating data between the virtual twin and the physical world. The minimum requirements, a virtual twin, the physical twin represented and a channel of live data, were met and successfully created using open-source GIS hosted with cloud-computing tools. The output was generic, the live data limited to global weather information rather than Internet of Things (IoT) sensor data. The spatial Digital Twin used open-source data except for the Digital Surface Model (DSM) representing Honiara's surface elevation. The investigation indicated that regions experiencing data scarcity can generate a city-model and implement a flow of updating data onto the model meets the minimum requirements of a spatial Digital Twin. The chapter's contribution conceptualises the minimum threshold for defining a spatial Digital Twin and its expected output and function. Further study

N. Lee (✉) · C. Sun · M. Wachowicz · D. Acharya
Geospatial Science, RMIT University, Melbourne, VIC 3000, Australia
e-mail: nicholas.lee@student.rmit.edu.au

C. Sun
e-mail: chayn.sun@rmit.edu.au

M. Wachowicz
e-mail: monica.wachowicz@rmit.edu.au

D. Acharya
e-mail: debaditya.acharya@rmit.edu.au

S. Ho
Department of Infrastructure Engineering, Faculty of Engineering and IT, University of Melbourne, Parkville, VIC 3052, Australia
e-mail: serene.ho@unimelb.edu.au

S. Pourroostaei Ardakani and A. Cheshmehzangi (eds.), *Digital Twin Computing for Urban Intelligence*, Urban Sustainability,
https://doi.org/10.1007/978-981-97-8483-7_3

topics highlighted is exploring what can be considered optimal features for a robust spatial Digital Twin, methods of obtaining highly detailed spatial data to improve its appearance, and sourcing different types of live information from areas that receive limited coverage. This will create a foundation for developing future spatial Digital Twins.

Keywords Spatial Digital Twin · Open-source solutions · GIS · Cloud computing · City-information model · Conceptualisation

1 Introduction

The chapter explores whether a spatial Digital Twin can be generated in a region that has a scarcity of spatial data—in this case, Honiara, the capital city of the Solomon Islands is used as a case study. The chapter focuses on the relationship between the theoretical definition of a spatial Digital Twin, its conceptual requirements and the output to illustrate the minimum elements achieved. The case study is used to explore the extent that open-source approaches can be leveraged to build a foundation for spatial Digital Twins and what are the minimum and optimal functions and attributes. An open-source approach allows for accessibility and replication of the methodology and encourages open-access and sharing of outputs to assist with future research work.

1.1 What Are Spatial Digital Twins and What Is Their Use?

A Digital Twin is defined as a realistic virtual replica of a physical object and its associated systems and processes [5]. The difference between a Digital Twin and city-model is that a Digital Twin is designed to reflect the real-time status of the object and system. The reflection is achieved by the Digital Twin being receptive to a continuous channel of updating information which can be from physical Internet of Things (IoT) sensors directly obtaining information and sending it back to the Digital Twin, or an Application Programming Interface (API) communicating information from another device. This gives the Digital Twin the ability to react to real-time data and self-update in response to real world changes from the information it receives. This creates a feedback loop between the real-world object and virtual twin [5]. Even if a city-model can display processes such as traffic in a city, it will only demonstrate data from a particular period rather than data based on the real-time state of the physical twin. Spatial Digital Twins reflect and represent phenomena and physical objects in a real-world geospatial context. These include real-world objects that can be represented as geospatial data such as buildings and roads and the processes that occur within the objects such as monitoring floods within a city or traffic in a town in real-time [2]. Spatial Digital Twins differentiate from other Digital Twins by representing real-world objects in a spatial context to monitor the state of objects and systems related to location-based information.

The application of spatial Digital Twins is using its self-updating capabilities to continuously monitor an object, area and its associated processes and system in real-time. An example is the City of Zurich in Switzerland which is a city-model that is consistently being updated to reflect any changes occurring in the real-world phenomena captured by the city's in-built data sensors [18]. Spatial Digital Twins have been commonly used for urban activity and studies. The prospect of spatial Digital Twins can be applied for assessing the impact and magnitude of activity within Honiara such as urban activity. The spatial Digital Twin of the City of Zurich in Switzerland goal is to respond to urban issues such as investigating how urban development can influence climate change within Zurich [18]. A spatial Digital Twin can assist decision-making to improve urban planning by providing an updated analysis of the area of representation. The real-time aspect provides an advantage over static alternatives which are limited to the timeframe the data was obtained in.

1.2 What Are the Current Research Gaps?

An identified research gap is that less developed countries do not get represented as a spatial Digital Twin or receive as much spatial analysis compared to richer countries. The Solomon Islands is considered the least developed country in the Pacific Region where the country relies heavily on outside aid [15]. The case studies of spatial analysis in Honiara investigated were all 2D analysis using static data. The studies assessed vulnerability of the population based on the presence and location of health system infrastructures [15]. A spatial Digital Twin can assist this case study by monitoring the distribution of people with illness and which areas has the highest population density of sick people. The spatial Digital Twin can indicate which health facilities are the closest to household areas and the current capacity of the facilities on whether they can accept more patients. This can assist the population more efficiently compared to a 2D static map application which can show the shortest route but does not indicate the current spread of illness or which health facilities can accept new patients.

A technical gap in the research is finding live data for data-scarce regions. The smart cities and spatial Digital Twins investigated in the literature uses IoT-based sensors in the city and citizen feedback obtained from local authorities to obtain data for the feedback loop. Docklands in Dublin, Ireland [22] and Zurich, Switzerland [18] have reliable continuous data channels as they integrate sensors into the cities to obtain local data. This allows the city-models to be considered spatial Digital Twins as it obtains live data and be continuously updated. Honiara has little spatial data and is not known to contain sensors in the city for obtaining local data. This means that obtaining live data will be difficult for Honiara.

Another technical gap is accessing live data and whether it is open-source and available for civilian use. Optimally, the data is captured from IoT sensors and automatically updates in real-time to receive accurate local information [7]. Although APIs can transmit real-time data, this information is dependent on what APIs are available. There is also whether the data is frequently updated and if its subject is relevant for the spatial Digital Twin's purpose. If the information does not frequently update automatically without user input, then the spatial Digital Twin would be considered a Digital city-model and not a twin [8].

1.3 Technical Objectives for the Lifecycle of a Foundational Spatial Digital Twin

The aim of this chapter is to develop a foundational spatial Digital Twin for Honiara, a city in the Pacific region, based on the conceptual definition and principles of Digital Twin technology. This involves exploring the necessary methodologies, tools, and data requirements to develop an accurate and functional digital replica of the city, especially in the context of data scarcity.

The chapter addresses several research questions to achieve this aim: Firstly, it investigates the conceptual requirements necessary to develop a 3D model as a spatial city-model Digital Twin, focusing on the fundamental principles and standards that define such a model. It then investigates why meeting these conceptual requirements is essential for qualifying the city-model as a Digital Twin, examining the specific criteria that set a Digital Twin apart from a basic 3D model. The chapter also evaluates the open-source data, programs, and automatically flowing live data available online that are suitable for the Digital Twin project, identifying resources that can be used to support the development of the Digital Twin. Finally, it explores how to create a Digital Twin using only open-source data and tools, outlining the steps required to generate the 3D city-model and develop it into a functioning web-based application.

The technical objectives are as follows:

- To gather and edit open-source spatial data to create the city information model and be exported into an interoperable 3D format.
- To create a back-end database and repository to store and manage the exported city information model.
- To identify a middleware component to connect and host the database onto the web and implement a form of self-updating data to the city information model.
- To publish the foundation spatial Digital Twin and self-updating data as a usable front-end application.
- To evaluate the prototype of foundation spatial Digital Twin.

The lifecycle of a foundational spatial Digital Twin involves several key stages. First, spatial data is collected using geospatial technologies, such as remote sensing and ground surveys. This data is then processed and turned into a digital model that accurately represents real-world assets and environments. The model can be continuously updated and validated to reflect real-time changes. Advanced tools and simulations can be used to predict future scenarios, aiding in proactive decision-making. The digital model, as a twin of its physical counterpart, is deployed on a platform that ensures it is accessible and can work with other systems (Duckham et al. 2024). Regular maintenance, including updates and system upgrades, is necessary to keep the Digital Twin current with technological advancements and user needs. Throughout its lifecycle, a foundational spatial Digital Twin can be used in applications such as asset management, operational efficiency, and strategic planning, making it a valuable tool in modern geospatial applications.

2 Literature Review

The literature review aims to identify the use of GIS in Honiara, open-source data in 3D modelling and how Digital Twins are used in industry activity. The literature review will investigate what has been currently conducted and achieved in the topics of interest along with highlighting gaps in the literature which has potential for further study. This review will identify the effectiveness of open-source data in 3D modelling along with current uses of GIS used in Honiara and Digital Twins in industry activity by investigating the extent of the results achieved and what limitations and issues were commonly faced.

2.1 Open-Source Data for 3D Modelling

Open-source data is online data that is freely accessible, available, and distributed and can be used with minimum restrictions by whoever uses the data [9]. An example of geospatial open-source data are shapefiles of buildings and road geometries that were obtained from measurements of a cadastral survey in Zurich [18]. Government data is considered more reliable than other online open data sources as the data is machine readable [9], receives regularly updates and uses measurements obtained from professional companies [18]. Another benefit of geospatial government data is that associated metadata is provided within the datasets themselves such as building permits in the associated area, crime report locations and transit routes [9]. Open data has been increasingly used in creating 3D spatial datasets for its availability and the aforementioned machine-readable nature allowing it to easily be integrated into 3D geospatial-based software and programs [13]. However, the literature has not gone into detail on whether a certain attribute or embedded value within the government data or how the data was measured makes it easier to convert into a 3D environment compared to other data types sourced from the same government body.

2.2 Open-Source Tools for 3D Modelling

Two reports have generated 3D models for investigating the area of representation by using open-source data and tools to build and analyse the model. The City of Zurich Digital Twin utilises Open Government Data for creating a spatial Digital Twin, demonstrating the capability and accuracy of Open Government Data [18]. Open-source programs can model 3D geographic data when used with Python scripts in the program Blender for creating and modelling the data to 3D, and PostGIS and PostgreSQL is used to create a web-based spatial database [19].

Open-source data is becoming prevalent and easier to access for cultural heritage data, but an issue with 3D modelling with solely open-source programs is that they might not always be easy to implement [16]. Despite success with Blender for 3D modelling, the journal has also identified areas of improvement such as improved Python scripting to optimise the code and calculation speed along with customised urban objects such as buildings and roads to improve drawing speed and realism of the model due to Blender being a 3D modelling tool rather than a program specifically designed for 3D GIS [19].

Another issue faced is that not all data files may be supported using open-source GIS programs. The coordinate system EPSG-900913 has been observed to not be supported in open-source GIS programs, the coordinate system otherwise allowing for faster and higher quality of integration of geo-web services [10]. However, the same article also observes that the EPSG-900913 is not supported in commercial-base GIS servers either. Compatibility and interoperability known as 'mashups' [10] are becoming more common with the prevalence of Web 2.0 and 3.0 technologies that encourage user-generated information, the journal article highlights the importance of open-source software continuously updating itself to allow for wider compatibility with different forms of data. For open-source programs for urban planning practices, an article details how some tools and programs might not be user-friendly without a background in programming or computers where 43% of programs explored requires a background in Python to be used to its full effectiveness [23].

Models made in three.js offers high interoperability and allows for efficient data federation as the three.js city-model can be viewed in any web browser that uses webGL [16]. Three.js offering a customisable library, high interoperability between web browsers and being a non-proprietary license demonstrates that three.js is an ideal open-source tool for generating 3D models as a web GIS application.

The literature has given mixed opinions on the effectiveness of using open-source tools for creating 3D spatial models. What can be extrapolated is that the issue with the software stems from the ease of usability by users and the programs capabilities for spatial modelling rather than the issue being that the open-source software cannot create the models at all. Contrary to issues with the open-source software, the literature covers how there is a push for users to use open-source data. This is prevalent within the City of Zurich example where the Digital Twin solely used open-source data to showcase how free-to-use data can be of a very high quality if it is properly detailed.

2.3 Principles of Developing Spatial Digital Twins and Their Application

A spatial Digital Twin may be defined by its conceptual definition but it should ideally provide a purpose. The Gemini Principles are guidelines for building a foundational definition of what a Digital Twin is, its three founding values and what is considered an appropriate level of detail. The three values of what a Digital Twin must have is a clear purpose, trustworthiness and an effective function [6]. The founding values provide a broad high-level scope of the quality expectation of information, data and purpose a Digital Twin needs to provide. These principles provide a connection between the theoretical definition of a Digital Twin and their application so that a theoretical Digital Twin can be well-defined in its purpose and function. The ANZLIC Spatial Information Council's Principles for Spatially Enabled Digital Twins of the Built and Natural Environment in Australia provides a hierarchical framework of how different subjects of data can be integrated together to provide advanced analytics from the spatial Digital Twin [3]. The framework aims to improve insights from the spatial Digital Twin, providing more efficient decision-making and improve the outcomes from the results and function of the spatial Digital Twins.

Spatial Digital Twins should have the capability of addressing a specific issue and focus on the issue of investigation if they are aligned with the Gemini Principles. The existing alternative is a smart-city Digital Twin that uses a multitude of urban information but does not address a specific issue or problem in high detail. A smart-city Digital Twin of Docklands in Dublin, Ireland has high visual detail and uses IoT data sources that include traffic data, moisture sensors in trees, rainfall and river levels [22]. The Digital Twin provides multiple purposes but appears to focus on obtaining a high number of different data sources rather than analysing the data to address an existing problem. The flooding simulation only raises the water level of the entire city without providing much detail on how this is calculated or captured. Little is discussed on any analysis of the captured data as the Digital Twin focuses on simulations and obtaining data. The Digital Twin of Docklands may be focused on citizen feedback but with no specific issue it aims to address, it lacks direction on what it should aim to achieve. Like the Digital Twin of Docklands, the spatial Digital Twin of Honiara should ideally address a specific issue.

Detailed city-models have been used extensively for generating spatial Digital Twins to represent physical features of a city. Compared to 3D models, spatial Digital Twins are used in conjunction with Web 2.0 and Web 3.0 technology so that the spatial data is continuously updated. The City of Zurich utilises a Digital Twin to assist with urban planning to tackle issues such as how activity will change due to climate change and demonstrating how phenomena will affect Zurich [18]. Digital Twins as a web-based application has advantages compared to static 3D models. Being a web application allows for continuous updating, whether it is from government data which regularly updates, live data obtained from sensors, or feedback from citizens who can identify any issues found within the spatial Digital Twin.

2.4 Digital Twins Live Information—Linking the Physical Counterpart to the Digital Twin

Spatial Digital Twins are linked through an IoT connection to link the physical counterpart to the Digital Twin. An IoT connection collects local data obtained from sensors from the real world to the virtual twin while the virtual twin returns rapid real-time data analytics and decisions made into the physical world [8]. A live data channel is the conceptual requirement of a spatial Digital Twin. The article indicates that data collected from IoT sensors and the Digital Twin returning analytics for real-time decision making is an optimal element present on a spatial Digital Twin.

Feedback loops are used in spatial Digital Twins to provide real-time data on processes that occur within the area of interest. How information is obtained into the feedback loops are from sensors that are distributed within a city, being utilised in the Digital Twin of Docklands, Dublin, Ireland [22]. The sensors used include smart trees moisture sensors, vehicles to report traffic to for scheduling traffic lights timing for optimising traffic flow, and wind sensors for identifying possible areas and designs for a wind farm. The Docklands Digital Twin utilises the 3D aspect for simulations to model situations that can occur within a city such as crowd simulations within the city and flood simulations to identify which areas will be flooded and to plan an evacuation route.

The effectiveness and cost of sensors and reliability of citizen data are taken into consideration to ensure that the information in the feedback loops is accurate and reliable. A journal article aimed to minimise the sensors used for cost efficiency and the number of feedback votes from citizens so that survey fatigue is reduced [1]. Although feedback loops were not mentioned, the City of Zurich Digital Twin utilises IoT sensors and feedback from citizens so that the Digital Twin can reflect the real-world processes based on user-input [18]. This is because the City of Zurich uses sensors and citizen feedback reported to the city authorities to optimise the smart city aspect of Zurich's Digital Twin.

3 Methodology

3.1 Framework and Conceptualisation for the Digital Twin

A Digital Twin conceptually consists of a physical twin, virtual twin and channel that communicates data and information between the two [17]. A conceptual framework of a spatial Digital Twin is showcased below in Fig. 1 which demonstrates the relationship between the real world and Digital Twin. The continuous flow of live data in the form of a feedback loop is what connects the physical twin and its virtual representation and differentiates the Digital Twin from a 3D model or simulation. The feedback loop incorporates live data to provide a reflection and analysis

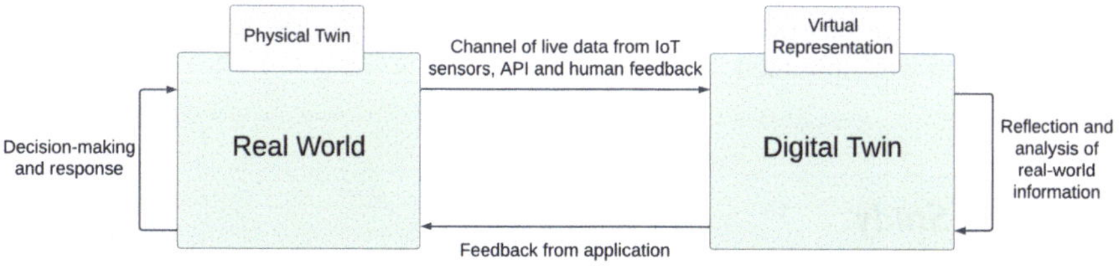

Fig. 1 Conceptual framework of a Digital Twin that demonstrates how a channel of live data creates the connection between the physical world to the virtual twin

of the information which is returned to the real world where decision-makers can be informed and respond to the new information from the Digital Twin.

By the theory of what is considered a Digital Twin, the spatial Digital Twin of Honiara created consisted of the real-world city, the virtual city-model created, and an ongoing flow of real-time data obtained from the physical world. These three criteria met the minimum requirements of what is conceptually considered a Digital Twin created using spatial city data. Figure 2 illustrates the workflow used to generate a Digital Twin based on conceptual theory. The framework was broken into five steps: (1) choosing a study area and determining whether it is feasible to create the Digital Twin based on area availability. (2) Data collection of the study area from open-source data resources. (3) Editing and manipulating data to make the initial model complete, detailed and interoperable where possible. (4) Installing and modifying model for an API to obtain live data. (5) Online repository creation so that the Digital Twin can be displayed on the web.

The framework aimed to make the Digital Twin detailed where possible but prioritised creating an interoperable real-time city-model over generating a highly detailed Digital Twin. This approach allowed the capabilities of open-data and programs to be explored and how well a Digital Twin could be created based on what was available and whether it met the theoretical definition of a Digital Twin. This approach utilised spatial data for the city-model creation, APIs for the continuous stream of

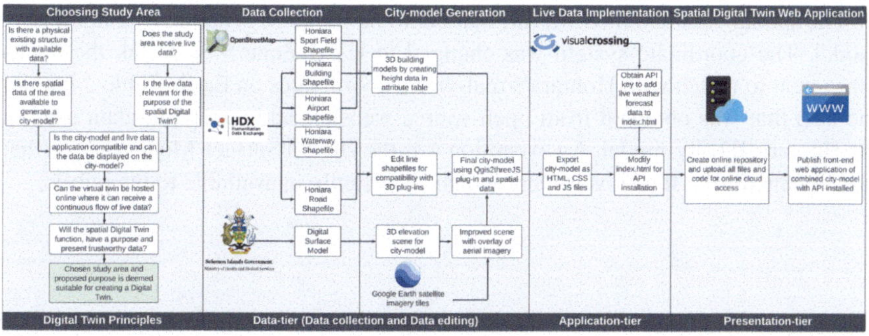

Fig. 2 Architectural framework of creating the Digital Twin of Honiara

live information and an open-source online repository and hosting program where the application was uploaded.

3.2 Case Study

Honiara is located on the Guadalcanal Island which is the largest of the Solomon Islands. Honiara is also considered the capital city of the Solomon Islands. The use of GIS in Honiara has few journal articles. The methodology of the use of GIS in Honiara only used GIS for spatial analysis and mapping on 2D planes with none of the literature utilising GIS for 3D analysis or model generation.

The GIS deliverables produced using Honiara data has been almost entirely distribution mapping and 2D analysis. Although customised data such as specialised spatial databases and automatic geo-referencing of malaria cases have been conducted and utilised, [11] the literature of Honiara's GIS activity is all 2D spatial analysis of solely distribution mapping on the topic of health and vulnerability assessment.

Honiara was chosen as the case study area because it is a data-scarce region. The case study's aim was to generate a spatial Digital Twin of Honiara using open-source spatial data and live information that is available and accessible. The fundamental spatial Digital Twin highlights the data-scarcity of Honiara based on the final output. The case study area of Honiara, located on the Solomon Islands was chosen to be the area of investigation, Fig. 3 illustrating its location relative to the Guadalcanal Island it is located on.

3.3 Open-Source Tools and Data Acquisition

The tools and data used to generate the spatial Digital Twin were all open-source. Table 1 lists the tools that were used.

The spatial data obtained was all that could be found to generate a detailed city-model. The coordinate system was changed to UTM Zone 57S to link the data's placement to match with Honiara's real-world coordinates on Earth. Table 2 lists all the data that was obtained from open-source websites and the spatial data used in the Honiara 3D city-model. An exception was the Digital Surface Model (DSM) that was obtained from the Government which is currently unavailable to the public.

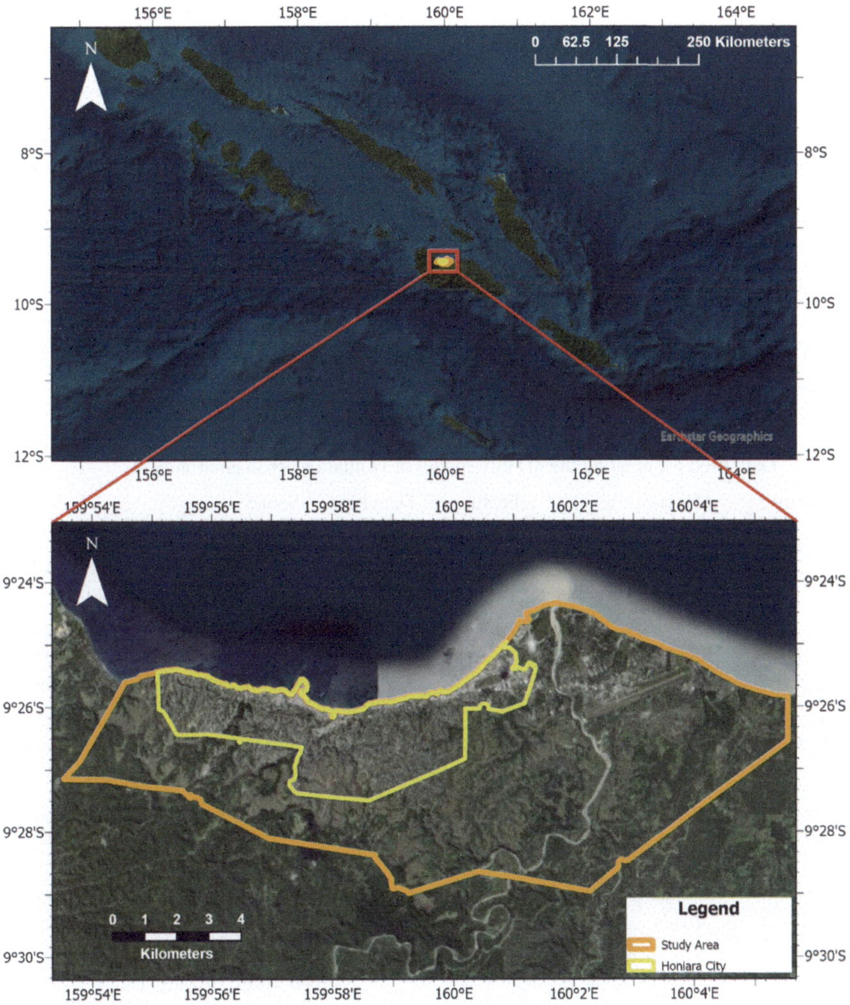

Fig. 3 Location of Honiara and locator map relative to the Guadalcanal island

3.4 *Generating the City-Model*

The DSM was used to create the elevation and surface of Honiara represented as a 3D scene. Unlike the other data, the DSM was obtained directly from the Solomons Islands government and is not available as an open-source dataset online.

The buildings and airport shapefiles had no height data provided since the data was only the building footprints. None of the Honiara spatial had any official height data available online. A set placeholder of values were implemented into all the building shapefiles attribute table to address this issue. This was so that the buildings have a value to use as its height/z-value when it gets exported as a city-model. The values

Table 1 Open-source tools used for creating the spatial Digital Twin

Program name	Tool function	Source
QGIS	Geographic Information System (GIS) used to edit spatial data	https://www.qgis.org/en/site/
QuickOSM	QGIS plugin used to download OpenStreetMap (OSM) data	https://plugins.qgis.org/plugins/QuickOSM/
Qgis2threejs	QGIS plugin used to preview and export spatial data into a 3D format	https://plugins.qgis.org/plugins/Qgis2threejs/
Visual Studio	Integrated Development Environment to edit code and script	https://visualstudio.microsoft.com/
GitHub	Platform to store, host and publish code onto a repository	https://github.com/

Table 2 Data used for creating the 3D city-model of Honiara's urban environment

Category	Dataset description	Feature type	Data type	Source
DSM	Raster dataset of the elevation in Honiara	Raster	GeoTIFF	Ministry of Health and Medical Services, Government of Solomon Islands
Satellite imagery	Satellite imagery overlay of 3D scene created	Raster	GeoTIFF	Google Earth Satellite imagery with no labels (XYZ Tiles from QGIS)
Buildings	Building footprints of the city	Polygon	Shapefile	https://data.humdata.org/dataset/hotosm_slb_buildings
Waterways	Dataset of all the waterways	Line	Shapefile	https://data.humdata.org/dataset/hotosm_slb_waterways
Roads	Dataset of all the roads	Line	Shapefile	https://data.humdata.org/dataset/hotosm_slb_roads
Airport	Footprint of the airport	Polygon	Shapefile	https://data.humdata.org/dataset/hotosm_slb_airports
Sport field	Footprints of sport fields and a sport centre	Polygon	Shapefile	OpenStreetMap export
Self-updating weather forecast	RESTful API that displays weather forecast	Live data	HTML/CSS/JS	https://github.com/visualcrossing/WeatherApi

applied were large enough to be visible on the DSM but not too high that it would cause the buildings to be too elongated or exaggerated.

The line datasets of the road and waterway data were placed flatly on the DSM itself rather than protruding from the DSM like the building shapefiles. This was achieved by preloading the road and waterway line features on the 2D surface rather than have them extruded as 3D objects like the buildings. Using QGIS's functions of

XYZ Tiles, Google Earth Satellite imagery was applied to overlay the scene so that the city-model's scene has realistic detail rather than being a mono-colored render.

Qgis2threejs was used to create the city-model as it utilises the open-source three.js JavaScript library. 3D technology is essential for developing spatial Digital Twins. After comparing various 3D development engines such as Unreal, Unity, Babylon.js, and three.js, it was found that three.js offers the lowest development cost. This makes three.js an attractive option for developing foundational spatial Digital Twins, allowing for efficient and cost-effective creation and maintenance of these complex digital models. The non-proprietary license emphasises openness of the city-model and its high interoperability between web browsers demonstrates the accessibility of models created in three.js.

The building shapefiles were placed as 3D objects using the plugin Qgis2threejs. The DSM was used as the elevation scene to represent the real-world terrain of Honiara. Vertical exaggeration applied to the DSM and building data where it was enough to illustrate the elevation and building heights but not too high that it would look inaccurate to how the buildings would look in the physical world. Exaggeration was applied as the default DSM generation did not provide appropriate visualisation on the 3D plane. The 3D building footprint shapefiles were added to create urban infrastructure of the city-model. The new height data column created on the building footprint data was used as the z-value height data so that the building shapefiles were generated in 3D. The sports field received little vertical exaggeration as it was needed to extrude from the DSM as it would otherwise sink and overlap with the DSM and not fully appear.

The final city-model consisted of the DSM vertically exaggerated with satellite imagery overlaid to create a detailed elevation surface. Waterway and road data was added as 2D data so that it overlays on top of the DSM while the sports field and city-model building data used the added height data as the z-values and its base were relative to the DSM. The final city-model was exported as an HTML file with the associated CSS and JavaScript files using the Qgis2threejs plugin. These files convert the display from Qgis2threejs into files that can be viewed as a basic city-model that can be zoomed and rotated with built-in functions, edited by opening the HTML, CSS and JavaScript code, and shared onto the web.

3.5 Weather Data Retrieval

An automatic channel of live self-updating data assists provides detail to the city-model and defines it as a spatial Digital Twin. A live data channel was installed and added to the city-model display in the form of an API. The API installed was a REST API from Visual Crossing that outputs forecasted weather data of Honiara. What is returned is maximum and minimum temperature, precipitation, and the expected weather conditions. Visual Crossing provided the API key while its GitHub repository had open-source code for the live weather forecast and a template to showcase the data. The number of days that can be forecasted is dependent on the vertical size

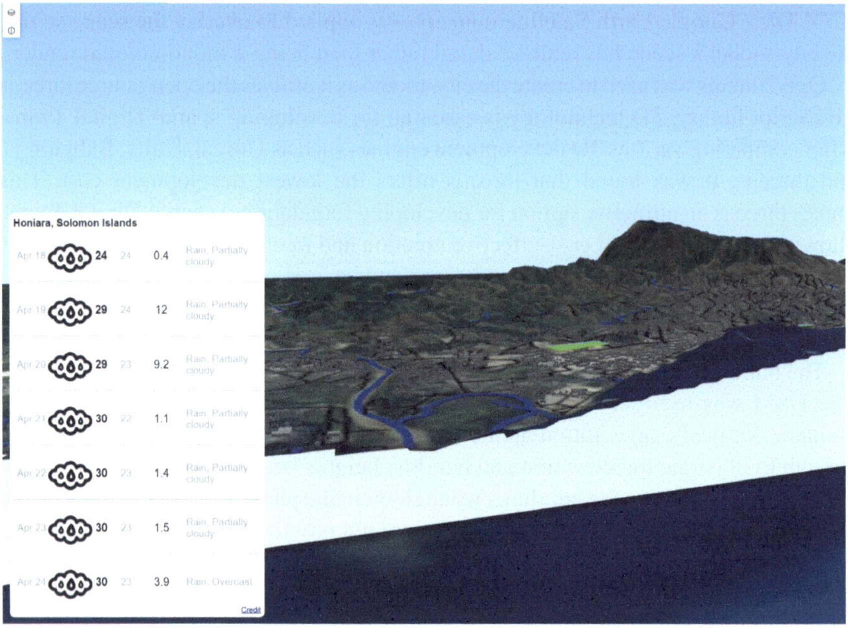

Fig. 4 Example of the weather API (bottom left corner) relative to the 3D city-model and the information it displays

of the display window in the provided template. The advantage of Visual Crossing is that it provides an API and a code template for displaying the data compared to other online APIs. Other weather APIs were an application that returns data but required coding knowledge to use the provided. The template and ability to be easily implemented onto HTML code was why Visual Crossing was chosen as the API.

The REST API was installed by editing the code of the city-model's HTML file using the output display code from Visual Crossing's GitHub repository [21]. The code was obtained and edited so that the display of the REST API did not interfere with the other existing elements within the city-model. This included editing the size of the weather output and ensuring that its placement did not overlap with the existing buttons in the model. An API key was obtained and put into the code as the key was needed to be able to obtain the live weather information. Figure 4 displays the placement of the REST API in the corner of the model where the size and placement were modified by editing the HTML and CSS of the REST API's code.

3.6 *Front-End Output and Hosting*

The city-model had live data implemented but required an online hosting server or repository to host and display the output as a web GIS application. The data files were all HTML, CSS and JavaScript files along with associated images needed for the weather API's display. The front-end web GIS application was hosted onto an online repository that can store the files together and be accessed on the web. The repository acted as a host that displays the HTML file as a web application that can be displayed and accessed to the public as a website. The chosen web repository was GitHub as it was able to upload the data files onto the web as open-source data along with displaying the HTML file as a functioning web application.

3.7 Web-Based Spatial Digital Twin Interface

The final spatial Digital Twin created was a basic city-model of Honiara. The application's function allows the user to view the Honiara city-model and weather forecast for the next seven days. It is possible to investigate the current distribution and density of buildings and the roads and waterways within Honiara relative to the area of interest within the elevated DSM.

The opacity of each layer can be changed so that the user can turn individual layers on and off as they please if they want to view one layer but remove the visual clutter of another. The user can zoom in and out by scrolling with the mouse wheel. Holding the left mouse button down while moving the mouse will allow rotations of the spatial Digital Twin.

The user can inspect the buildings and structures that are present throughout Honiara relative to the area of interest. The API has successfully been implemented where live weather data is provided on the application. As seen in Fig. 5, the distribution of urban structures and the built environment can be assessed compared to their placement on the elevated surface throughout the DSM's elevation and surface features.

4 Discussions

4.1 Discussion of Results and Limitations

The output met the minimum requirements of a spatial Digital Twin. The virtual twin was the city-model and scene generated and the live data loop was the weather API added into the application. Conceptually, this fulfills the three requirements of a Digital Twin of a virtual twin representing a physical twin with the presence of continuous data to self-update [5]. Honiara city is the physical twin, the generated city-model and scene is the virtual twin and the API that provides a seven-day weather forecast for Honiara is the continuous data.

Conceptually, the spatial Digital Twin aligns with the Gemini Principles where the data used follows interoperability and federation through its three-tier architecture

Fig. 5 Final display of the spatial Digital Twin of Honiara. The spatial digital twin can be viewed while the table on the bottom left is a live weather forecast

nature. The spatial Digital Twin can evolve with new technologies as it is developed on a three.js platform, enabling high customisation and interoperability with any webGL-based web browser. The purpose is creating the foundation and development for future spatial Digital Twins in data-scarce regions to promote investigation and case studies of cities that receive little research. The Gemini Principle of trustworthiness is not completely aligned due to the data quality. However, the open-source approach means that the spatial Digital Twin's methodology promotes openness, allowing developers to replicate the methodology and access to the same tools used.

However, the virtual twin represented is a generic model of Honiara city, using shapefiles of the buildings with a low level of detail (LOD). The live information communicated between the city-model and the physical world is weather forecast data obtained through an API. This contrasts from the Digital Twin representing Docklands in Dublin, Ireland which uses sensors and citizen feedback to provide local data [22]. The API obtained was basic and sourced from global weather data rather than being specialised data obtained and exclusive to Honiara itself. A limitation was the availability of open-source spatial data in Honiara. There is the possibility to obtain higher quality data from private sources. The DSM obtained from the Government of Solomon Islands, Ministry of Health and Medical Service was the only dataset obtained that could add detail to the spatial Digital Twin which was not open-source.

The spatial Digital Twin's appearance was generic as it was dependent on the data availability. However, the investigation was whether a spatial Digital Twin could be generated for a data-scarce region. The result was a basic Honiara city-model placed on a government sourced DSM that is currently unavailable to the public. Although the DSM goes against the study of using open-source data, using it was optional as it was to add elevation to the city-model rather than being a mandatory component. A limitation was the height accuracy of the buildings as the spatial data did not provide the height data.

A technical limitation with the open-source tools was what was used to generate the city-model: Qgis2threejs. There were issues identified with placing the building shapefiles onto the DSM as there was no method to individually move the buildings around. This caused some of the buildings to be "merged" halfway through the DSM or floating above the surface. This occurred with all shapefiles that were placed as a shapefile, including the line features of waterways and roads. Figure 6 illustrates this issue which is a prototype version of the Digital Twin where the road and waterway features were placed as 3D line features relative to the 3D scene, causing them to "float" over the scene.

The final spatial Digital Twin had the building shapefiles and sports fields be applied as 3D objects while the roads and waterways were overlaid as a 2D file. The roads appeared thicker and blurry compared to using the line shapefiles as 3D overlays, but the line features all had less issues in appearance. Compare Figs. 6 and 7 on the benefits and disadvantages between displaying the line features as a 3D extrusion (above) and 2D overlay (below).

Cesium, a 3D geospatial program had the building footprint data with a height value. Unlike areas with higher spatial accuracy such as Melbourne City where the height data varied, Honiara's buildings all shared the same value of "3". This highlights an inaccuracy of the building data present on Cesium. Houses located in rural areas shared the same height as the airport and the building footprints did

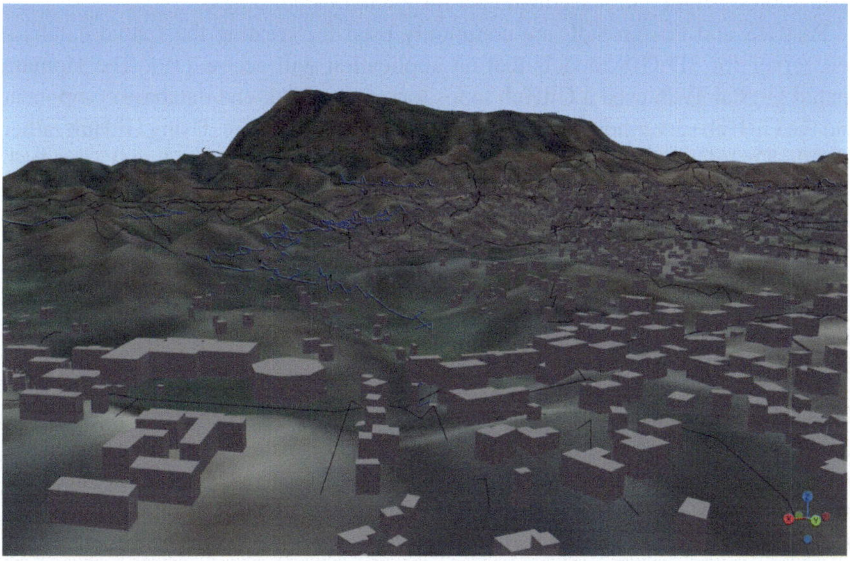

Fig. 6 Issue faced with placing the building, road and waterway shapefiles relative to the DSM on a prototype version of the spatial Digital Twin. Note the floating and merged buildings, roads (black lines) and waterways (blue lines)

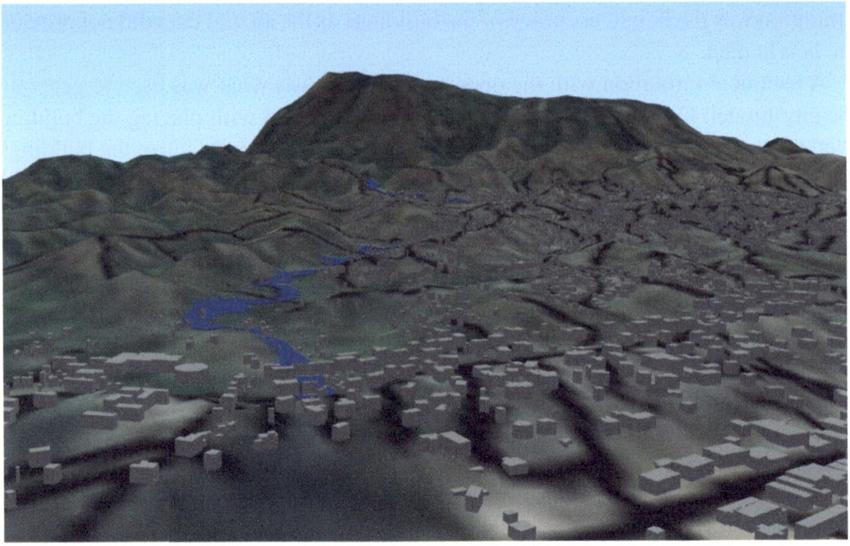

Fig. 7 Final Digital Twin where only the buildings are relative to the DSM while the roads and waterways were overlaid on the DSM as a 2D file

not always match up with what was shown on the satellite imagery [12]. Cesium's Honiara data indicates that there are currently no reliable sources to obtain accurate open-source height and placement data for Honiara's buildings.

PostGIS and PostgreSQL are commonly used for creating the spatial database and server for 3D GIS models that an application will access [19]. The Honiara spatial Digital Twin used a GitHub repository as the back-end database component and the GitHub server hosting the repository is the middleware. Using GitHub rather than the OpenGeoSuite is not common but has been conducted. A Digital Twin Web concept called "Twinbase" also implements GitHub to act as a Digital Twin server [4], similar to what this spatial Digital Twin used.

4.2 Tools Considered

The Digital Twin used a GitHub repository and Git database as the middleware server and backend database software respectively. Using GitHub-based open-source software is different compared to other spatial Digital Twins and web GIS applications which primarily use PostgreSQL with the PostGIS extension as the database component and GeoServer as the application/server component. A prototype version of the spatial Digital Twin used a PostgreSQL database using the PostGIS extension and has uploaded the data onto GeoServer where a connection has been secured.

GeoServer was considered as the middleware application but there was no long-term free option available for the server to be continuously running. An option was a Microsoft Azure virtual machine to have GeoServer running but that would not be an open-source solution as it requires a purchase to use.

4.3 Potential Data Sources

There are a multitude of other open data sources to add further detail onto the spatial Digital Twin. These sources and types range from historical data, highly detailed spatial data to create an informative city-model and different APIs to provide either accurate live information or different topics of data to provide a wider range of real-time detail and analysis.

Historical data implemented into the spatial Digital Twin's database can be used as a validation method based on the issue the spatial Digital Twin aims to address. A study has used historical flood information to validate flood risk maps, achieving an accuracy of 83% when sorting the flooded points into risk classes [14]. This demonstrates how historical data can be used in future spatial Digital Twins to investigate disasters or phenomena by comparing the conditions of past occurrences to predict future scenarios by identifying common patterns and trends.

Detailed spatial data adds more information to the spatial Digital Twin. Open-source government data is considered the most reliable source compared to other online open data sources as the information is free to use, has minimal restrictions for using it and is in a machine readable format [9]. Government spatial data is also regularly updated and the measurements were obtained from professional companies [18]. Countries with more frequently updated open-source data available will have spatial data with higher accuracy in its measurements and detail in its metadata. The open-source data that was used was from OpenStreetMap (OSM) and the Humanitarian Data Exchange. The drawbacks are that OSM data is sourced from volunteers which cannot guarantee the accuracy of the data and may either update infrequently or stop altogether. The Humanitarian Data Exchange sources its information from OpenStreetMap for most of its Honiara-based data. The Honiara datasets update every month according to the Humanitarian Data Exchange. However, there is no guarantee on data quality as it is done by an OpenStreetMap volunteer team rather than a government or private organisation. The metadata embedded had few details where it is entirely textual information such as river name and what each airport shapefile represented such as a control tower or the helipad.

Sources of live information that are primarily used on a Digital Twin are IoT sensors embedded into the city or APIs. IoT sensors collect specialised information. These can include moisture sensors for trees, wind sensors for areas that a wind farm can be built in [22], temperature, smoke and images [20]. Information obtained from IoT sensors provides a direct connection between the physical and virtual world compared to APIs that are obtained from other computers. An idea of what is defined as a Digital Twin considered IoT sensors on the physical twin as a crucial requirement

[17]. Utilising IoT sensors to obtain a direct channel of communication between the physical and virtual twin provides a higher accuracy of time-critical information. This enables Digital Twins to be used for functions requiring high data velocity such as emergency services of disaster management. The IoT sensors alleviates the issue that an API may face such as obtaining specific information that is currently unavailable from an API. Additionally, the sensor can provide another form of data communication for the same information to compare effectiveness of the two sources and be used as a back-up data flow in case the API connection is compromised. There is no known presence of IoT sensors within Honiara to provide an accurate and self-updating stream of information. These sensors would benefit the Honiara Digital Twin by providing a direct channel of communication and real-time data, allowing the Digital Twin to be used to address a particular purpose.

Countries and locations with a high presence of real-time government-sourced data can obtain APIs to generate detailed spatial Digital Twins for a specific purpose. Data from the Victorian Government contains APIs for specific activity such as road disruptions, freeway travel time and agricultural soil data. These APIs use real-time information that is more accurate to Victoria compared to the Visual Crossing API that uses real-time global data to obtain approximate location data by entering the city-name or coordinates. The APIs by the Victorian Government provide the frequency of update as well, being 5 calls per minute on average. These APIs can be integrated into other Digital Twins to provide more accurate and location-specific live information, updating at a regular frequency.

4.4 Future Research

Future research benefits from this study by investigating whether other data-scarce regions can create a spatial Digital Twin based on the conceptual definition. Honiara is an area that receives few spatial studies or investigations. This study exemplifies that a spatial Digital Twin can be created if the requirements of a virtual city-model and live data installation is possible. To improve future spatial Digital Twins, investigating how to improve the data's representation such as higher resolution satellite imagery and accuracy of the building dimensions assists with providing higher detail. The spatial Digital Twin developed demonstrates the current scarcity of spatial data in Honiara, highlighting urban areas that would benefit from an improved detail of spatial data. Another area of study for future research is whether variations in the systems architecture used to share the application can provide further benefits for hosting and maintaining the Digital Twin.

GitHub, which was used in this study, can be an open-source tool for future spatial Digital Twins as it is an online repository that hosts data and publish pages. Future projects also use GitHub like this study and Twinbase in case alternate methods such as PostGIS and PostgreSQL are not appropriate for future spatial Digital Twins.

Although the DSM was not open-source, many areas now have coverage of open-sourced LiDAR data which are available online which can help with future studies aiming to investigate a similar topic.

Future applications using three.js for Digital Twins are expansive and promising. In urban planning, three.js can be used to create dynamic, real-time 3D models of cities, helping planners visualise and simulate urban growth and infrastructure changes. In the construction industry, it can facilitate the development of detailed building models, allowing for enhanced project visualisation, planning, and collaboration among stakeholders. In environmental monitoring, three.js can help create interactive models that simulate and predict environmental changes, aiding in disaster management and sustainability efforts. Additionally, three.js can be leveraged in smart city projects to integrate various data streams into a cohesive 3D model, enabling better decision-making and management of urban resources. Its low development cost and flexibility make three.js a valuable tool for a wide range of future applications in the realm of Digital Twins.

5 Conclusion

A foundational spatial Digital Twin was successfully created based on its conceptual definition of a virtual twin representing a physical object that responds to real-world changes using a channel of live data. Using this conceptual definition, it is possible to create a spatial Digital Twin of a data-scarce region if there exists spatial data of the area and the ability to implement a feedback loop of live data into the city-model's application. Despite meeting the conceptual definition, the spatial Digital Twin contained basic information and was limited by the data availability of the area. The final product was a basic city-model where the live data was globally sourced weather forecasts rather than local information due to the data scarcity of Honiara. The minimum requirements were met but not the optimal requirements of obtaining IoT sensor data and returning detailed data analytics or real-time decisions [8]. These findings can be used to support future research by investigating whether other areas with similarly low amounts of spatial data can have a Digital Twin generated, provided there is the availability and presence of relevant live information. The results can be improved by increasing the level of detail and implementing live information. Investigating other methods for obtaining live information would be beneficial so that more information can be communicated between the physical object and the virtual model for future Digital Twins.

The theoretical contribution achieved is developing a spatial Digital Twin by its conceptual definition of a physical twin, virtual twin and a channel of live data between the twins. The chapter's theoretical contribution is creating a foundation for data-scarce areas to develop their own spatial Digital Twins based on the conceptual definition. The case study area, Honiara, has had few spatial studies conducted where the chapter demonstrates that a data-scarce and less developed region can have a spatial Digital Twin generated as a web application.

The technical contribution achieved is pioneering the development of spatial Digital Twins in data-scarce regions using open-source solutions. This chapter can achieve generating Honiara as a spatial Digital Twin where its open-source nature allows users to replicate the methodology and build upon it with any data they have obtained themselves. The three-tiered distribution system used provides a foundation for how to share the spatial Digital Twin as a web application.

Acknowledgements I would like to thank and acknowledge the Government of Solomon Islands, Ministry of Health and Medical Services for Lidar data for providing the basis of the DSM.

Appendix: Data and Codes Availability

Repository: https://github.com/nicholaslee264/HoniaraDT2024.

Web link for Digital Twin: https://nicholaslee264.github.io/HoniaraDT2024/.

References

1. Abdelrahman MM, Miller C (2022) Targeting occupant feedback using digital twins: adaptive spatial–temporal thermal preference sampling to optimize personal comfort models. Build Environ 218:109090. https://doi.org/10.1016/j.buildenv.2022.109090
2. Ali ME, Cheema MA, Hashem T, Ulhaq A, Muhammad Ali B (2023) Enabling spatial digital twins: technologies, challenges, and future research directions. arXiv.org. https://doi.org/10.48550/arxiv.2306.06600
3. ANZLIC (2019) Principles for spatially enabled digital twins of the built and natural environment in Australia. https://www.anzlic.gov.au/resources/principles-spatially-enabled-digital-twins-built-and-natural-environment-australia
4. Autiosalo J, Siegel J, Tammi K (2021) Twin base: open-source server software for the digital twin web. IEEE Access 9:140779–140798. https://doi.org/10.1109/ACCESS.2021.3119487
5. Batty M (2018) Digital twins. Environ Plan B, Urban Anal City Sci 45(5):817–820. https://doi.org/10.1177/2399808318796416
6. CDBB (2018) The Gemini principles. In: U. O. Cambridge (ed), Digital Twin Hub. https://digitaltwinhub.co.uk/download/the-gemini-principles/
7. Dani AAH, Supangkat SH, Lubis FF, Nugraha IGBB, Kinanda R, Rizkia I (2023) Development of a smart city platform based on digital twin technology for monitoring and supporting decision-making. Sustain (Basel, Switzerland) 15(18):14002. https://doi.org/10.3390/su151814002
8. Fuller A, Fan Z, Day C, Barlow C (2020) Digital twin: enabling technologies, challenges and open research. IEEE Access 8:108952–108971. https://doi.org/10.1109/ACCESS.2020.2998358
9. Johnson PA, Sieber R, Scassa T, Stephens M, Robinson P (2017) The cost(s) of geospatial open data. Trans GIS 21(3):434–445. https://doi.org/10.1111/tgis.12283
10. Karnatak HC, Shukla R, Sharma VK, Murthy YVS, Bhanumurthy V (2012) Spatial mashup technology and real time data integration in geo-web application using open source GIS—a case study for disaster management. Geocarto Int 27(6):499–514. https://doi.org/10.1080/10106049.2011.650651

11. Kelly GC, Hale E, Donald W, Batarii W, Bugoro H, Nausien J, Smale J, Palmer K, Bobogare A, Taleo G, Vallely A, Tanner M, Vestergaard LS, Clements ACA (2013) A high-resolution geospatial surveillance-response system for malaria elimination in Solomon Islands and Vanuatu. Malar J 12(1):108–108. https://doi.org/10.1186/1475-2875-12-108
12. Lee N, Sun Q, Ho S, Wachowicz M, Acharya D (2024) Rapid development of a spatial digital twins using open-source solutions. In: ISPRS technical commission IV symposium 2024, Fremantle, Perth, Australia
13. Mobasheri A, Pirotti F, Agugiaro G (2020) Open-source geospatial tools and technologies for urban and environmental studies. Open Geospatial Data, Softw Stand 5(1). https://doi.org/10.1186/s40965-020-00078-2
14. Mohammadi M, Darabi H, Mirchooli F, Bakhshaee A, Torabi Haghighi A (2021) Flood risk mapping and crop-water loss modeling using water footprint analysis in agricultural watershed, Northern Iran. Nat Hazards (Dordrecht) 105(2):2007–2025. https://doi.org/10.1007/s11069-020-04387-w
15. Natuzzi ES, Joshua C, Shortus M, Reubin R, Dalipanda T, Ferran K, Aumua A, Brodine S (2016) Defining population health vulnerability following an extreme weather event in an urban Pacific Island environment: Honiara, Solomon Islands. Am J Trop Med Hyg 95(2):307–314. https://doi.org/10.4269/ajtmh.16-0177
16. Nishanbaev I, Champion E, McMeekin DA (2021) A web GIS-based integration of 3D digital models with linked open data for cultural heritage exploration. ISPRS Int J Geo Inf 10(10):684. https://doi.org/10.3390/ijgi10100684
17. Piroumian V (2023) Making digital twins work. Computer (Long Beach, Calif.) 56(1):42–51. https://doi.org/10.1109/MC.2022.3206101
18. Schrotter G, Hürzeler C (2020) The Digital Twin of the City of Zurich for Urban Planning. J Photogramm Remote Sens Geoinf Sci 88(1):99–112. https://doi.org/10.1007/s41064-020-00092-2
19. Scianna A (2013) Building 3D GIS data models using open source software. Appl Geomat 5(2):119–132. https://doi.org/10.1007/s12518-013-0099-3
20. Shi J, Pan Z, Jiang L, Zhai X (2023) An ontology-based methodology to establish city information model of digital twin city by merging BIM. GIS and IoT. Adv Eng Inform 57:102114. https://doi.org/10.1016/j.aei.2023.102114
21. VisualCrossing (2019) Visual crossing weather API repository. https://github.com/visualcrossing/WeatherApi
22. White G, Zink A, Codecá L, Clarke S (2021) A digital twin smart city for citizen feedback. Cities 110:103064. https://doi.org/10.1016/j.cities.2020.103064
23. Yap W, Janssen P, Biljecki F (2022) Free and open source urbanism: software for urban planning practice. Comput Environ Urban Syst 96:101825. https://doi.org/10.1016/j.compenvurbsys.2022.101825

Fusion of Digital Twin, Internet-of-Things and Artificial Intelligence for Urban Intelligence

Houman Zarrabi and M. R. Doost Mohammadian

Abstract This chapter explores the fusion of Digital Technologies (DT), Internet of Things (IoT), and Artificial Intelligence (AI) for enhancing urban intelligence and decision-making processes. It investigates how these technologies are integrated to improve various aspects of urban life, including sustainability, resilience, democratic governance, and community empowerment. Drawing on a comprehensive review of the literature, case studies, and theoretical frameworks, the chapter provides insights into the potential benefits, challenges, and implications of adopting integrated smart city solutions. The integration of DT, IoT, and AI enables cities to collect, analyze, and act on vast amounts of data in real-time, leading into more informed and proactive decision-making. For example, AI-driven predictive analytics help optimize resource allocation, enhance service delivery, and improve infrastructure management, contributing to the efficient and sustainable operation of urban systems. Moreover, IoT sensors and connected devices enable cities to monitor environmental conditions, traffic flows, and energy consumption, providing valuable insights for urban planning and policy formulation. Furthermore, the chapter discusses how integrated technologies contribute to promoting digital inclusion, civic engagement, and democratic governance in cities. By providing platforms for citizen participation, feedback, and collaboration, cities can foster transparency, accountability, and responsiveness in government. Citizen-centric innovation approaches, such as co-creation workshops, empower residents to contribute their expertise and local knowledge to the design and implementation of smart city initiatives, ensuring that solutions are tailored to community needs and priorities.

Keywords Digital twin · Internet of Things · Artificial intelligence · Information fusion · Urban intelligence

---

H. Zarrabi (✉)
Iran Telecom Research Center, Tehran, Iran
e-mail: h.zarrabi@itrc.ac.ir

M. R. Doost Mohammadian
Sharif University of Technology, Tehran, Iran

Semnan University, Semnan, Iran

S. Pourroostaei Ardakani and A. Cheshmehzangi (eds.), *Digital Twin Computing for Urban Intelligence*, Urban Sustainability,
https://doi.org/10.1007/978-981-97-8483-7_4

1 Introduction

Urban areas around the world are facing unprecedented challenges due to rapid population growth, resource constraints, and environmental degradation. In response to these challenges, cities are increasingly turning to digital transformation (DT), Internet of Things (IoT), and artificial intelligence (AI) to enhance urban intelligence and decision-making processes. This chapter explores how the fusion of these technologies is reshaping urban landscapes, improving efficiency, and fostering sustainability.

Urban Digital transformation involves the integration of digital technologies into all aspects of business and society, fundamentally altering operations and processes. In urban contexts, digital transformation manifests in various forms, including the digitization of public services, the deployment of smart infrastructure, and the adoption of data-driven decision-making practices. For example, cities are implementing open data platforms to make government data accessible to the public, enabling transparency and accountability in governance.

The Internet of Things (IoT) refers to the network of interconnected devices that collect and exchange data in real-time. In urban environments, IoT sensors are deployed across infrastructure systems such as transportation, energy, and water to monitor performance and detect anomalies. By leveraging IoT data, cities can optimize resource allocation, reduce downtime, and enhance service delivery. For instance, smart traffic management systems use IoT sensors to monitor traffic flow and adjust signal timings dynamically, alleviating congestion and improving mobility.

Artificial intelligence (AI) encompasses algorithms and techniques that enable computers to perform tasks that traditionally require human intelligence. In urban contexts, AI is used for predictive analytics, enabling cities to forecast future trends and anticipate events. For example, AI algorithms analyze historical crime data to identify high-risk areas and allocate police resources effectively. Additionally, AI-powered energy management systems optimize energy consumption in buildings based on occupancy patterns and weather forecasts, leading to cost savings and reduced environmental impact.

The integration of DT, IoT, and AI holds immense potential for improving urban intelligence and decision-making. By combining data from diverse sources and leveraging advanced analytics, cities can gain deeper insights into urban dynamics, identify emerging trends, and respond proactively to challenges. Moreover, integrated technologies enable cities to optimize resource allocation, enhance service delivery, and improve quality of life for residents. In the following sections, we will explore how these technologies are being integrated in various urban domains to address complex challenges and create more sustainable, resilient, and livable cities.

The methodological framework embraces the idea of a digital twin, demonstrated in the following Fig. 1. A digital twin creates a virtual representation of the physical urban city to enhance its efficiency. This virtual model relies on extensive, continuous, real-time data collected from various aspects. By utilizing this data, digital replicas

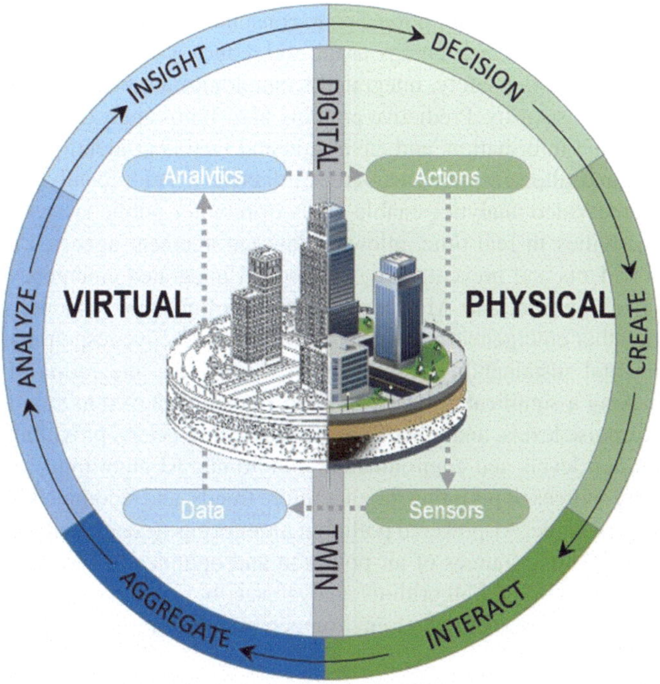

Fig. 1 Urban intelligence and its digital twin counterpart [1]

of the city are constructed, offering valuable insights into urban performance. These insights can drive real-world interventions like modifications in urban planning or transportation systems.

Verticals and Use Cases in Urban Intelligence

One of the key areas where the fusion of DT, IoT, and AI is making a significant impact is transportation. Smart transportation systems leverage IoT sensors, AI algorithms, and data analytics to optimize traffic flow, reduce congestion, and improve safety. For example, intelligent traffic management systems use real-time data from IoT sensors to detect traffic incidents and congestion, allowing traffic signals to be adjusted dynamically to optimize traffic flow. Additionally, AI-powered predictive analytics enable cities to forecast demand for transportation services and allocate resources efficiently, leading to reduced waiting times and improved accessibility for residents.

In the energy sector, integrated technologies are being deployed to enhance efficiency and sustainability. Smart grids leverage IoT sensors and AI algorithms to optimize the distribution of electricity, reduce energy losses, and accommodate renewable energy sources. For instance, AI algorithms analyze data from IoT sensors to predict electricity demand patterns and optimize the operation of grid infrastructure, leading to cost savings and reduced carbon emissions. Moreover, IoT-enabled smart meters provide real-time energy consumption data to consumers, empowering them to make informed decisions about their energy usage and reduce waste.

In the realm of public safety, integrated technologies are being used to prevent crime and enhance security. Predictive policing algorithms analyze historical crime data, demographic information, and environmental factors to identify areas at high risk of crime and allocate police resources effectively. Additionally, IoT sensors and AI-powered video analytics enable cities to monitor public spaces and detect suspicious activities in real-time, allowing law enforcement agencies to respond quickly to incidents and prevent crime. Moreover, integrated emergency response systems use IoT sensors and AI algorithms to coordinate the response to natural disasters and other emergencies, ensuring a swift and effective response to crises.

Environmental sustainability is another area where the integration of DT, IoT, and AI is making a significant impact. IoT sensors are deployed to monitor air and water quality, noise levels, and other environmental parameters, providing real-time data on pollution levels and environmental conditions. AI algorithms analyze this data to identify sources of pollution, predict future trends, and recommend mitigation strategies. For example, AI-powered pollution monitoring systems analyze data from IoT sensors to identify sources of air pollution and optimize traffic flow to reduce emissions. Additionally, AI algorithms analyze satellite imagery to monitor deforestation and identify areas at risk of environmental degradation, enabling proactive conservation efforts.

In the healthcare sector, integrated technologies are being used to improve patient outcomes and service delivery. IoT devices such as wearable sensors and remote monitoring devices enable continuous monitoring of patients' vital signs and health metrics, allowing healthcare providers to detect health issues early and intervene proactively. Additionally, AI algorithms analyze medical imaging data such as X-rays and MRI scans to assist in diagnosis and treatment planning, leading to more accurate and timely interventions. Moreover, telemedicine platforms leverage DT infrastructure to provide remote consultations and healthcare services to underserved populations, improving access to healthcare and reducing disparities in healthcare delivery.

Water management is another area where integrated technologies are making a significant impact. IoT sensors are deployed to monitor water quality, detect leaks, and optimize water distribution networks. AI algorithms analyze this data to predict water demand, optimize water usage, and identify opportunities for conservation. For example, AI-powered leak detection systems analyze data from IoT sensors to identify leaks in water pipelines and prioritize maintenance activities, leading to reduced water losses and lower operational costs. Additionally, IoT-enabled smart irrigation systems optimize water usage in agriculture by monitoring soil moisture levels and weather forecasts, leading to improved crop yields and water efficiency.

Intelligent Urban Planning and Governance

In urban planning, integrated technologies are enabling data-driven decision-making and more sustainable development. Geographic information systems (GIS) leverage DT infrastructure to collect, store, and analyze spatial data on land use, transportation, and infrastructure. AI algorithms analyze this data to identify patterns, predict future trends, and inform urban planning decisions. For example, AI-powered land use models analyze demographic data and economic indicators to forecast demand for housing and commercial space, informing decisions about zoning and development. Additionally, IoT sensors collect real-time data on transportation patterns, air quality, and other urban parameters, enabling planners to monitor the impact of development projects and adjust plans accordingly to ensure sustainability and livability.

Citizen engagement is essential for the success of smart city initiatives and the integration of DT, IoT, and AI. By involving residents in decision-making processes and providing opportunities for feedback and participation, cities can ensure that smart city projects reflect the needs and priorities of the community. For example, participatory budgeting initiatives allow residents to propose and vote on projects to allocate municipal funds, empowering them to shape the future of their neighborhoods. Additionally, online platforms and mobile apps enable residents to report issues such as potholes, graffiti, and street lighting outages, facilitating quick response and resolution by city authorities. Moreover, community workshops and public forums provide opportunities for residents to learn about smart city technologies and provide input on their implementation, fostering a sense of ownership and accountability in the community.

Effective governance and policy frameworks are essential for guiding the responsible deployment and use of integrated technologies in urban contexts. Policymakers must balance the benefits of innovation with considerations of privacy, security, and equity, ensuring that smart city initiatives prioritize the public interest. For example, data privacy regulations such as the General Data Protection Regulation (GDPR) in the European Union establish guidelines for the collection, storage, and use of personal data, protecting the rights of individuals while enabling the responsible use of data for public purposes. Additionally, ethical guidelines for AI development and deployment, such as the IEEE Ethically Aligned Design framework, provide principles and best practices for ensuring that AI systems are fair, transparent, and accountable, minimizing the risks of bias, discrimination, and harm.

Integrated technologies have the potential to drive economic development and innovation in urban areas, creating new opportunities for businesses, entrepreneurs, and workers. By fostering a supportive ecosystem for innovation and entrepreneurship, cities can attract investment, talent, and ideas, stimulating economic growth and job creation. For example, innovation districts such as Silicon Valley in California and Station F in Paris provide physical spaces and resources for startups, research institutions, and investors to collaborate and develop new technologies and business models. Additionally, smart city initiatives such as open data platforms and innovation challenges provide opportunities for businesses to access data, develop new products and services, and contribute to the urban innovation ecosystem.

In conclusion, the fusion of DT, IoT, and AI is transforming urban landscapes, enabling cities to become smarter, more efficient, and more sustainable. By integrating data from diverse sources and leveraging advanced analytics, cities can gain deeper insights into urban dynamics, optimize resource allocation, and improve decision-making processes. Moreover, integrated technologies enable cities to address complex challenges such as transportation congestion, energy inefficiency, and public safety, creating more livable and resilient urban environments. However, realizing the full potential of integrated technologies requires addressing challenges such as data privacy, security, and digital inclusion, as well as fostering collaboration and engagement among stakeholders. Through concerted efforts and strategic investments, cities can harness the power of DT, IoT, and AI to build more inclusive, equitable, and prosperous urban futures.

2 Barriers to Adoption

Despite the promising prospects of integrated DT, IoT, and AI for urban intelligence, several barriers hinder widespread adoption and implementation. These barriers encompass technological, organizational, regulatory, and socio-economic dimensions, posing formidable challenges to cities seeking to harness the full potential of smart technologies.

One major barrier is the lack of interoperability and compatibility between disparate systems and platforms. As cities invest in a multitude of IoT devices and AI solutions from different vendors, ensuring seamless integration and data exchange becomes increasingly complex. Interoperability standards and protocols are needed to facilitate interoperability, enabling cities to deploy integrated solutions that span multiple domains and stakeholders.

Another barrier is the shortage of skilled talent and expertise in emerging technologies such as AI and data science. Many cities struggle to attract and retain qualified professionals with the requisite skills to design, deploy, and maintain complex DT, IoT, and AI systems. Investments in workforce development, education, and training are essential to build a talent pipeline capable of driving innovation and sustainability in urban contexts.

One of the key benefits of integrated DT, IoT, and AI for urban intelligence is the potential to enhance sustainability and resilience in cities. By leveraging real-time data insights and predictive analytics, cities can optimize resource allocation, improve energy efficiency, and mitigate environmental risks.

For example, AI-driven energy management systems enable cities to dynamically adjust energy consumption patterns based on demand forecasts, weather conditions, and pricing signals. By optimizing the operation of lighting, heating, and cooling systems in buildings, cities can reduce energy waste, lower utility costs, and curb carbon emissions, contributing to climate mitigation efforts.

Furthermore, IoT sensors and AI algorithms play a critical role in enhancing urban resilience to natural disasters and extreme weather events. By monitoring environmental conditions, detecting anomalies, and issuing timely warnings, cities can minimize the impact of disasters on infrastructure, public safety, and human health. Moreover, AI-powered predictive modeling enables cities to anticipate future risks and prioritize investments in resilience-building measures, such as flood defenses, green infrastructure, and emergency response capabilities.

As cities embrace the transformative potential of integrated DT, IoT, and AI, it is imperative to address ethical considerations and ensure that smart city initiatives uphold fundamental principles of fairness, transparency, and accountability. Ethical challenges encompass a wide range of issues, including data privacy, algorithmic bias, and digital exclusion, which have profound implications for social justice and human rights.

One ethical concern is the equitable distribution of benefits and risks associated with smart city technologies. While DT, IoT, and AI have the potential to improve urban services and quality of life for residents, there is a risk that marginalized communities may be left behind or disproportionately burdened by negative externalities such as surveillance, discrimination, and loss of autonomy. Ensuring inclusive participation and representation in the design and implementation of smart city initiatives is essential to address these disparities and promote social equity.

Another ethical consideration is the responsible use of urban data and AI algorithms. As cities collect vast amounts of data from IoT sensors, mobile devices, and public infrastructure, concerns arise regarding data privacy, consent, and surveillance. Transparent data governance frameworks, privacy-preserving technologies, and community-driven oversight mechanisms are needed to safeguard individual rights and freedoms while harnessing the power of data for public good.

Effective governance and policy frameworks are essential to guide the responsible deployment and utilization of integrated DT, IoT, and AI in urban contexts. Policymakers must strike a delicate balance between fostering innovation and protecting public interests, ensuring that smart city initiatives align with democratic values, human rights, and sustainability goals.

One key policy area is data governance and privacy protection. Cities need robust legal frameworks and regulatory mechanisms to govern the collection, storage, sharing, and use of urban data, safeguarding individual privacy rights and preventing misuse or abuse of personal information. Moreover, transparency and accountability measures are essential to build trust and confidence among citizens, enabling informed consent and meaningful participation in decision-making processes.

Given the multifaceted nature of urban challenges and the interdependence of stakeholders, collaborative governance approaches are essential to foster effective partnerships and collective action. By bringing together government agencies, private sector actors, civil society organizations, and academic institutions, cities can leverage diverse perspectives, resources, and expertise to address complex urban problems holistically.

One example of collaborative governance in action is the formation of public–private partnerships (PPPs) to finance and implement smart city projects. PPPs enable cities to leverage private sector investments and expertise while retaining control over public assets and service delivery. By sharing risks and rewards, PPPs can accelerate innovation, reduce costs, and enhance the scalability and sustainability of smart city solutions.

Moreover, citizen engagement and participatory decision-making processes are essential for ensuring that smart city initiatives reflect the needs, preferences, and aspirations of local communities. Through open forums, online platforms, and community workshops, cities can solicit feedback, co-design solutions, and co-create knowledge with residents, fostering a sense of ownership and empowerment among stakeholders.

Integrated DT, IoT, and AI play a crucial role in enhancing the resilience of urban infrastructure to shocks and stresses, ranging from natural disasters and climate change to technological failures and social disruptions. By leveraging real-time data insights and predictive analytics, cities can anticipate, prepare for, and respond to emerging threats in a proactive and adaptive manner.

For example, IoT sensors embedded in critical infrastructure assets such as bridges, water pipelines, and power grids enable cities to monitor structural health, detect anomalies, and prioritize maintenance activities. By detecting early warning signs of deterioration or failure, cities can prevent catastrophic failures, minimize service disruptions, and extend the lifespan of infrastructure assets.

Furthermore, AI-powered predictive modeling enables cities to assess and mitigate risks associated with extreme weather events, such as floods, hurricanes, and heatwaves. By analyzing historical data, weather forecasts, and climate projections, cities can identify vulnerable areas, develop risk mitigation strategies, and allocate resources effectively to protect lives and property.

As cities embrace the transformative potential of integrated DT, IoT, and AI, it is essential to ensure that smart city initiatives are inclusive and accessible to all residents, regardless of age, gender, ability, or socioeconomic status. Digital inclusion encompasses a wide range of considerations, including affordable broadband access, user-friendly interfaces, and tailored services for diverse user needs.

In conclusion, the fusion of DT, IoT, and AI holds immense promise for enhancing urban intelligence and decision-making in cities around the world. By leveraging real-time data insights, predictive analytics, and citizen engagement principles, cities can address complex urban challenges and improve quality of life for residents. However, realizing the full potential of integrated urban intelligence requires concerted efforts to overcome barriers to adoption, address ethical considerations, and ensure inclusive and accessible smart city initiatives. Through collaborative governance, resilient infrastructure, and equitable policies, cities can build more sustainable, resilient, and inclusive urban futures for generations to come.

3 Privacy and Security as Major Concerns

The integration of DT, IoT, and AI in urban contexts raises significant concerns regarding privacy and security. As cities collect and analyze vast amounts of data from sensors, devices, and infrastructure, there is a risk of unauthorized access, data breaches, and misuse of personal information. Moreover, the proliferation of surveillance technologies such as facial recognition systems and smart cameras raises questions about individual privacy rights and civil liberties.

To address these concerns, cities must implement robust data protection measures, encryption protocols, and access controls to safeguard sensitive information. Additionally, transparent and accountable data governance frameworks are essential to ensure that data collection and usage practices are lawful, ethical, and aligned with privacy regulations. Citizen awareness and engagement are also crucial for fostering trust and confidence in smart city initiatives, enabling residents to make informed decisions about data sharing and privacy preferences.

One of the primary goals of integrated DT, IoT, and AI in urban contexts is to deliver citizen-centric services that meet the diverse needs and preferences of residents. By leveraging real-time data insights and predictive analytics, cities can personalize services and tailor interventions to individual or community-level requirements. For example, AI-powered chat-bots and virtual assistants enable residents to access information, report issues, and request services through intuitive and conversational interfaces, enhancing user experience and satisfaction.

Moreover, IoT-enabled smart infrastructure such as connected streetlights, waste bins, and parking meters can optimize resource allocation, improve service delivery, and enhance quality of life for residents. For instance, smart parking systems use IoT sensors and AI algorithms to detect available parking spaces in real-time, guiding drivers to vacant spots and reducing traffic congestion. Similarly, smart waste management systems optimize waste collection routes based on fill-level sensors, minimizing operational costs and environmental impact.

Integrated DT, IoT, and AI enable cities to make data-driven decisions across various domains, from transportation and energy to public safety and healthcare. By harnessing real-time data insights and predictive analytics, cities can optimize resource allocation, improve service delivery, and respond proactively to emerging challenges.

For example, AI-powered predictive maintenance systems analyze data from IoT sensors to detect equipment failures before they occur, enabling preventive maintenance actions to be taken to minimize downtime and repair costs. Similarly, predictive analytics models forecast demand for public services such as transportation and healthcare, allowing cities to allocate resources efficiently and mitigate service disruptions.

Moreover, AI algorithms analyze vast amounts of urban data to identify patterns, trends, and correlations that may not be apparent to human analysts. By uncovering hidden insights and actionable intelligence, cities can make informed decisions that drive innovation, efficiency, and sustainability in urban development.

The implementation of integrated DT, IoT, and AI in urban contexts often requires significant upfront investments in technology infrastructure, human capital, and organizational capacity. However, many cities face budgetary constraints and competing priorities, making it challenging to secure funding for smart city initiatives.

To overcome these challenges, cities are exploring innovative financing mechanisms such as public–private partnerships (PPPs), impact investing, and crowdfunding. PPPs enable cities to leverage private sector investments and expertise while retaining control over public assets and service delivery. Impact investing involves the allocation of capital to projects that generate social and environmental benefits alongside financial returns, attracting funding from socially conscious investors and philanthropic organizations. Crowdfunding platforms enable cities to raise funds for specific smart city projects through contributions from residents, businesses, and community groups, fostering a sense of ownership and engagement in the community.

As cities embrace the benefits of integrated DT, IoT, and AI, it is essential to ensure that all residents have access to and benefit from these technologies, regardless of their socio-economic status, geographic location, or digital literacy. Digital inclusion encompasses a wide range of initiatives aimed at bridging the digital divide and empowering marginalized communities to participate in the digital economy.

To promote digital inclusion and equity, cities must invest in education and capacity building initiatives that enhance digital literacy, skills development, and access to technology resources. By providing training programs, workshops, and community centers, cities can empower residents with the knowledge and skills to navigate the digital landscape, access online services, and participate in the digital economy.

Moreover, cities can leverage partnerships with educational institutions, non-profit organizations, and private sector firms to expand access to technology resources and support lifelong learning opportunities. For example, coding boot camps, STEM education programs, and digital literacy workshops equip residents with the technical skills and competencies needed to thrive in the digital age.

Integrated DT, IoT, and AI thrive in collaborative ecosystems characterized by open innovation, knowledge sharing, and cross-sectoral collaboration. By fostering partnerships between government agencies, academic institutions, industry players, and civil society organizations, cities can leverage diverse perspectives, resources, and expertise to address complex urban challenges and drive innovation.

In conclusion, the fusion of DT, IoT, and AI has the potential to revolutionize urban intelligence and decision-making, enabling cities to become smarter, more efficient, and more sustainable. By integrating data from diverse sources and leveraging advanced analytics, cities can gain deeper insights into urban dynamics, optimize resource allocation, and improve service delivery. Moreover, integrated technologies enable cities to address complex challenges such as transportation congestion, energy inefficiency, and public safety, creating more livable and resilient urban environments.

However, realizing the full potential of integrated urban intelligence requires addressing barriers to adoption, ethical considerations, and digital inclusion challenges. Through collaborative governance, innovative financing mechanisms, and investment in education and capacity building, cities can build inclusive, equitable, and prosperous urban futures for all residents.

4 Sustainable Urban Development

Integrated DT, IoT, and AI play a crucial role in advancing sustainable urban development by optimizing resource use, reducing environmental impact, and enhancing resilience to climate change. By harnessing real-time data insights and predictive analytics, cities can make informed decisions that promote economic prosperity, social equity, and environmental stewardship.

For example, AI-driven energy management systems optimize energy consumption in buildings, transportation, and industry, leading to cost savings and reduced carbon emissions. Similarly, IoT-enabled smart grids improve the efficiency and reliability of electricity distribution, integrating renewable energy sources and storage technologies to promote clean and resilient energy systems.

Moreover, AI-powered predictive modeling enables cities to assess and mitigate risks associated with climate change, such as extreme weather events, sea-level rise, and heatwaves. By developing adaptation strategies and investing in resilient infrastructure, cities can minimize the impact of climate-related disasters on public health, infrastructure, and economic productivity.

Integrated DT, IoT, and AI contribute to community resilience and social cohesion by fostering collaboration, empowerment, and collective action among residents. By providing platforms for civic engagement, knowledge sharing, and mutual support, cities can build stronger, more resilient communities that are better equipped to withstand and recover from shocks and stresses.

For example, digital platforms such as social media, online forums, and mobile apps enable residents to connect with each other, share information, and organize community-led initiatives. During emergencies, these platforms can facilitate real-time communication and coordination among residents, first responders, and local authorities, enabling swift and effective response efforts.

Moreover, IoT-enabled community resilience projects, such as neighborhood sensors networks and community gardens, empower residents to monitor environmental conditions, address local concerns, and build social capital. By fostering trust, collaboration, and social cohesion, cities can enhance community resilience and promote equitable development for all residents.

Integrated DT, IoT, and AI enable citizen-centric innovation by empowering residents to participate in the co-creation and co-design of smart city solutions. By providing opportunities for citizen engagement, feedback, and collaboration, cities can harness the creativity, expertise, and local knowledge of residents to address pressing urban challenges and improve quality of life.

For example, participatory design workshops, hackathons, and innovation challenges enable residents to propose ideas, develop prototypes, and pilot new technologies in partnership with city officials, researchers, and entrepreneurs. By involving residents in the design and implementation of smart city initiatives, cities can ensure that solutions are tailored to local needs, preferences, and priorities, leading to greater acceptance and adoption.

Moreover, citizen science initiatives, such as air quality monitoring projects and biodiversity surveys, enable residents to collect and analyze data on environmental conditions, contributing valuable insights to urban planning and decision-making processes. By democratizing access to data and knowledge, cities can empower residents to become active participants in the co-production of urban intelligence and resilience.

Integrated DT, IoT, and AI have the potential to transform democratic governance and civic engagement by enhancing transparency, accountability, and participation in decision-making processes. By leveraging digital technologies, cities can expand access to information, facilitate public deliberation, and strengthen democratic institutions, fostering a culture of trust, openness, and collaboration between government and citizens.

For example, open data platforms, such as data portals and transparency websites, enable cities to publish government data in accessible and machine-readable formats, empowering residents, journalists, and researchers to analyze and visualize information, hold government accountable, and advocate for policy change. Moreover, digital tools such as e-voting systems and participatory budgeting platforms enable residents to vote on policy decisions, allocate public funds, and engage in direct democracy processes, enhancing civic participation and political representation.

Furthermore, IoT-enabled civic engagement projects, such as smart city sensors networks and crowd-sourced mapping initiatives, enable residents to collect, share, and analyze data on urban conditions, contributing valuable insights to urban planning and decision-making processes. By harnessing the collective intelligence and creativity of residents, cities can address complex urban challenges more effectively and build resilient, inclusive, and democratic communities.

Building public trust and ensuring ethical governance are essential for the success of integrated DT, IoT, and AI in urban contexts. As cities deploy smart technologies to collect, analyze, and act on vast amounts of data, it is crucial to uphold principles of fairness, transparency, and accountability in the use of data and algorithms.

To build public trust, cities must adopt transparent and accountable data governance frameworks that govern the collection, storage, sharing, and use of urban data. Privacy-preserving technologies, encryption protocols, and access controls are essential to protect sensitive information and prevent unauthorized access or misuse of data. Moreover, cities must ensure that AI algorithms are transparent, explainable, and accountable, enabling residents to understand how decisions are made and challenge biased or discriminatory outcomes.

Additionally, cities must engage residents in ethical discussions and deliberative processes to co-create ethical guidelines and principles for the responsible use of DT, IoT, and AI in urban contexts. By involving stakeholders in decision-making processes and soliciting input on ethical considerations, cities can build consensus, foster trust, and ensure that smart city initiatives uphold fundamental values and human rights.

Responsible innovation and risk management are essential for ensuring the safe and ethical deployment of integrated DT, IoT, and AI in urban contexts. As cities experiment with new technologies and applications, it is crucial to assess and mitigate potential risks and unintended consequences, such as privacy violations, algorithmic bias, and social inequalities.

One approach to responsible innovation is the adoption of precautionary principles and risk assessment frameworks that anticipate and mitigate potential harms associated with smart city technologies. By conducting rigorous impact assessments and scenario analyses, cities can identify risks, evaluate trade-offs, and develop strategies to minimize negative impacts on society and the environment.

Moreover, cities must establish mechanisms for ongoing monitoring, evaluation, and adaptation of smart city initiatives to ensure that they remain aligned with public interests and values. Transparency and accountability measures, such as public reporting, independent audits, and citizen oversight, are essential to build public trust and confidence in smart city governance.

In conclusion, the fusion of DT, IoT, and AI holds immense promise for transforming urban intelligence and decision-making, enabling cities to become smarter, more resilient, and more inclusive. By integrating data from diverse sources and leveraging advanced analytics, cities can gain deeper insights into urban dynamics, optimize resource allocation, and improve service delivery. Moreover, integrated technologies enable cities to address complex challenges such as climate change, social inequality, and democratic governance, creating more sustainable, equitable, and democratic urban futures.

However, realizing the full potential of integrated urban intelligence requires addressing ethical considerations, promoting digital inclusion, and fostering public trust and participation. Through collaborative governance, responsible innovation, and investment in education and capacity building, cities can build resilient, inclusive, and prosperous urban ecosystems that benefit all residents.

5 Promoting Digital Inclusion and Equity

As cities embrace the potential of integrated DT, IoT, and AI, it is essential to prioritize digital inclusion and equity to ensure that all residents can access and benefit from smart city initiatives. Digital inclusion encompasses a range of initiatives aimed at bridging the digital divide and empowering marginalized communities to participate fully in the digital economy.

For example, cities can invest in programs that provide affordable internet access, digital literacy training, and access to technology devices for low-income households. Community centers and public libraries can serve as hubs for digital inclusion activities, offering computer classes, technology workshops, and assistance with online services. Additionally, cities can partner with local organizations and businesses to provide tailored support and resources to underserved communities, addressing barriers such as language, culture, and disability.

Empowering citizens with data literacy skills is essential for promoting informed decision-making and civic engagement in the era of integrated urban intelligence. Data literacy encompasses the ability to understand, interpret, and critically evaluate data, enabling individuals to make sense of complex information and participate effectively in public discourse and decision-making processes.

To promote data literacy, cities can invest in educational initiatives that teach residents how to collect, analyze, and visualize data relevant to their communities. School curricula can integrate data literacy concepts and skills across disciplines, from mathematics and science to social studies and civics. Moreover, adult education programs, community workshops, and online tutorials can provide opportunities for lifelong learning and skill development in data literacy.

Integrated DT, IoT, and AI have the potential to enhance civic participation and democratic governance by enabling residents to engage more effectively in public decision-making processes. By providing platforms for information sharing, deliberation, and collaboration, cities can foster a culture of transparency, accountability, and responsiveness in government.

For example, online platforms such as e-participation portals and civic engagement apps enable residents to access government information, submit feedback, and participate in consultations from the convenience of their smartphones or computers. Moreover, digital tools such as interactive maps, data dashboards, and budget simulators enable residents to explore complex issues, visualize data, and understand the impact of policy decisions on their communities.

Empowering communities through co-creation and co-design processes is essential for ensuring that smart city initiatives reflect the diverse needs, preferences, and priorities of residents. By involving community members in the design and implementation of projects, cities can harness local knowledge, creativity, and expertise to develop solutions that are contextually appropriate, culturally sensitive, and socially inclusive.

For example, participatory design workshops and design sprints bring together residents, planners, architects, and other stakeholders to generate ideas, prototype solutions, and co-create urban interventions. By fostering collaboration and consensus-building, these processes enable communities to take ownership of the development process and ensure that projects are aligned with their values and aspirations.

Moreover, co-design approaches enable cities to address systemic inequalities and power imbalances by centering the voices and experiences of marginalized communities in decision-making processes. By prioritizing equity, diversity, and inclusion, cities can build trust, foster social cohesion, and promote shared ownership of urban development initiatives.

Building collaborative ecosystems for innovation is essential for fostering a culture of open innovation, knowledge sharing, and cross-sectoral collaboration in smart cities. By bringing together government agencies, academic institutions, industry players, and civil society organizations, cities can leverage diverse perspectives, resources, and expertise to address complex urban challenges and drive innovation.

For example, innovation hubs, technology incubators, and co-working spaces provide physical spaces and resources for startups, entrepreneurs, and researchers to collaborate, experiment, and develop new solutions. By facilitating serendipitous encounters and interdisciplinary collaborations, these spaces foster creativity, accelerate innovation, and promote knowledge exchange.

Moreover, cities can establish innovation districts and living labs that serve as testbeds for new technologies and urban interventions. By providing real-world environments for experimentation and evaluation, these initiatives enable cities to pilot innovative solutions, gather feedback from residents, and iterate on designs before scaling up or replicating successful models.

6 Conclusions

In conclusion, the fusion of DT, IoT, and AI has the potential to revolutionize urban intelligence and decision-making, enabling cities to become smarter, more inclusive, and more resilient. By integrating data from diverse sources and leveraging advanced analytics, cities can gain deeper insights into urban dynamics, optimize resource allocation, and improve service delivery. Moreover, integrated technologies enable cities to address complex challenges such as social inequality, environmental sustainability, and democratic governance, creating more livable and equitable urban environments for all residents.

However, realizing the full potential of integrated urban intelligence requires addressing barriers to adoption, promoting digital inclusion, and fostering collaborative governance. Through strategic investments in education, capacity building, and community empowerment, cities can build resilient, inclusive, and prosperous urban ecosystems that benefit all residents.

References

1. Petrova-Antonova D, Ilieva S (2019) Methodological framework for digital transition and performance assessment of smart cities. In: 2019 4th international conference on smart and sustainable technologies (SpliTech), pp 1–6
2. Smith A (2019) The role of digital transformation in urban governance. J Urban Technol 21(3):345–362
3. Johnson B et al (2020) Leveraging IoT for smart transportation systems: case study of city X. IEEE Trans Intell Transp Syst 15(4):789–802
4. Chen C et al (2018) Applications of artificial intelligence in urban planning: a review. J Urban Plan Dev 144(1):05017011
5. Lee S et al (2021) Smart cities: concepts, technologies, and challenges. Wiley
6. Anthopoulos LG (2015) Understanding smart cities: a roadmap for city and urban development. J Knowl Econ 6(3):720–738
7. Deakin M, Al Waer H (2011) From intelligent to smart cities. Intell Build Int 3(3):140–152
8. Komninos N (2016) The age of intelligent cities: smart environments and innovation-for-all strategies. Routledge
9. Anthopoulos LG (2017) Understanding the smart city domain: a literature review. Smart Cities 1(1):6–25

10. Albino V et al (2015) Smart cities: definitions, dimensions, performance, and initiatives. J Urban Technol 22(1):3–21
11. Angelidou M (2017) Smart cities: a conjuncture of four forces. Cities 63:128–139
12. Batty M (2013) The new science of cities. MIT Press
13. Deakin M (2019) Smart cities: governing, modelling, and analysing the transition. Routledge
14. Evans-Cowley JS (2017) Smart cities in the USA. Smart Cities 1(1):104–120
15. Nam T, Pardo TA (2020) Advances in smart city development: a systematic literature review. Sustain Cities Soc 55:102022
16. Smith J, Johnson L (2022) Implementing high-performance work systems for enhanced organizational performance: a contemporary perspective. J Organ Excell 8(1):45–62
17. Hollands RG (2015) Critical interventions into the corporate smart city. Camb J Reg Econ Soc 8(1):61–77
18. Hollands RG (2018) Urban planning and the crisis of the inner city: the future of the inner city. Routledge
19. Mitchell WJ (1999) E-topia: urban life, Jim—but not as we know it. MIT Press
20. Townsend AM (2013) Smart cities: big data, civic hackers, and the quest for a new utopia. WW Norton & Company
21. Alshuwaikhat HM et al (2012) Smart cities: a critical review of the literature. Sustainabilitiy 5(12):951–987
22. Angelidou M (2015) Smart city policies: a spatial approach. Cities 47:80–91
23. Anthopoulos LG (2018) Understanding smart cities: a framework based on enabling technologies and citizen-centric factors. Int J Electron Govern 10(1):4–19
24. Caragliu A et al (2009) Smart cities in Europe. J Urban Technol 16(2):65–82
25. Hollands RG (2019) Critical perspectives on the smart city. J Urban Technol 26(2):1–6
26. Kitchin R (2014) The real-time city? Big data and smart urbanism. GeoJournal 79(1):1–14
27. Komninos N (2013) The age of intelligent cities: smart environments and innovation-for-all strategies. J Knowl Econ 4(2):163–185
28. Lacity MC et al (2016) Robotics, artificial intelligence, and service automation. MIS Q Exec 15(4):245–257
29. Nam T et al (2015) Is Seoul a smart city? Urban Stud 52(15):1–18
30. Albino V et al (2015) Understanding smart cities: an integrative framework. Intell Build Int 7(4):184–202
31. Angelidou M (2015) The role of smart city characteristics in the plans of fifteen cities. J Urban Technol 22(4):3–28
32. Anthopoulos LG (2017) The role of smart city characteristics in the plans of fifteen cities. J Urban Technol 24(2):3–28
33. Kim H, Lee S, Park H (2021) Integrating smart city characteristics into urban development plans: comparative analysis of twenty global cities. Sustain Cities Soc 65:102598
34. Deakin M (2016) Smart city governance: a comparative assessment of Edinburgh, San Francisco, and Singapore. Int J Urban Reg Res 40(3):618–635
35. Evans-Cowley JS (2019) Smart city governance: a comparative assessment of Edinburgh, San Francisco, and Singapore. J Urban Technol 21(3):345–362

Digital Twin Technologies

Building Resilient Smart Cities: The Role of Digital Twins and Generative AI in Disaster Management Strategy

Hooman Razavi, Omid Titidezh, Ali Asgary, and Hossein Bonakdari

Abstract The advent of Digital Twin (DT) technology represents a significant milestone in the evolution of smart city management, introducing virtual models and data-driven simulations that enhance our understanding, planning, and management of urban environments. Generative Artificial Intelligence (GenAI) integration enriches digital twins by boosting their predictive capabilities and simulating more realistic and interactive scenarios. By leveraging generative algorithms, digital twins can create synthetic data, to simulate a wide range of potential outcomes. These additional capabilities enable more accurate modeling of complex systems, predicting variations and potential issues. Moreover, Generative AI can contribute to creating high-quality simulations, improving the accuracy and reliability of digital twins in representing real-world environments and processes. This chapter provides an exploration of the synergies between GenAI and digital twins in the context of disaster management and smart cities. The chapter begins with an introduction, offering a contextual background integrating generative AI and digital twins for simulating disaster and emergency scenarios. It then delves into the foundations of generative AI, discussing its principles, applications, and success stories across various domains, with a specific focus on its relevance to urban disaster management. The subsequent sections clarify the evolution of digital twins and their pivotal role in predicting and mitigating disasters and emergency response. The chapter then navigates the intersection between generative AI and digital twins and the functionalities that GenAI brings to improving the simulation process, providing examples of successful

H. Razavi
School of Engineering Science, Tecnologico de Monterrey, Puebla, Mexico

O. Titidezh
School of Architecture, Building and Civil Engineering, Loughborough University, Loughborough, UK

A. Asgary
School of Administrative Studies, Faculty of Liberal Arts and Professional Studies, York University, Toronto, ON, Canada

H. Bonakdari (✉)
Department of Civil Engineering, University of Ottawa, Ottawa, ON, Canada
e-mail: hbonakda@uottawa.ca

S. Pourroostaei Ardakani and A. Cheshmehzangi (eds.), *Digital Twin Computing for Urban Intelligence*, Urban Sustainability,
https://doi.org/10.1007/978-981-97-8483-7_5

integrations while addressing potential challenges and proposing solutions. Furthermore, it explores how smart cities contribute to disaster resilience, detailing the technologies and strategies employed for disaster preparedness. The chapter concludes with an in-depth analysis of specific GenAI-enhanced digital twins' applications in disaster management, anticipating future challenges and developments in the field, and emphasizing emerging trends and potential directions.

Keywords Digital twin · Data synthesis · Disaster and emergency management · Generative AI · Scenario generation · Smart cities · Urban intelligence · Urban resilience

1 Introduction

Cities are intricate ecosystems that are rapidly evolving, mirroring the increasing digitization seen among citizens. Projections suggest that by 2025, the global internet user base will reach 6.2 billion, with connected devices surpassing 100 billion [3]. This surge in connected devices will generate approximately 180 ZB of data annually, a fivefold increase from current levels [65]. City managers are under increasing pressure to operate more efficiently as urban digital infrastructure evolves rapidly. Simultaneously, they must utilize networks of devices and data to cater to the needs of digital natives [56].

The rapid digitization of urban centers, driven by the rapid expansion of internet users and interconnected devices, underscores the pressing need for new smart city initiatives [31]. With urban populations on the rise and technological advancements steering change, smart cities offer a viable solution to effectively manage resources, enhance infrastructure, and elevate residents' quality of life [21]. By harnessing data from interconnected devices, smart city initiatives facilitate informed decision-making, streamline service delivery, and foster more resilient and sustainable cities and urban development [64]. As cities transform into intricate digital ecosystems, the emergence of smart cities transcends mere responses to technological shifts [18]; it becomes a strategic imperative for shaping the cities of the future [30]. As depicted in Fig. 1, the projected global revenue for smart city technologies, products, and services is expected to reach 165 billion U.S. $by 2028. This revenue encompasses incomes generated by all companies offering technologies and products utilizing information and data technologies to create added value within the urban environments. This includes entities providing essential infrastructure like cloud computing and connectivity technologies, enabling cities to leverage these advancements more effectively.

Digital twins, initially introduced by Grieves in 2002 for product lifecycle management, emerged as a key technological trend for addressing power system challenges [29]. DT holds promise in offering solutions for issues related to the optimal management, operation, control, and security of cyber-physical systems [71]. Defined as a digital model of a physical entity, DT accurately reflects its physical behavior

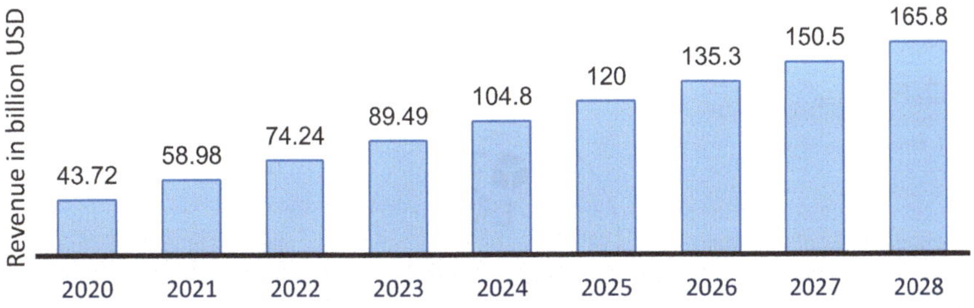

Fig. 1 Revenue generated by companies in the global smart city (*Source* Statista [60])

through real-time data interactions and platforms [63]. Until recently, the Digital Twin concept has emerged as a significant player in urban development, particularly in smart cities, offering innovative avenues to enhance environmental performance [32]. As a virtual representation of a physical system updated through information exchange between physical and virtual realms, DT has gained traction due to its potential to revolutionize how smart city infrastructures are managed and planned. This surge in interest is attributed to insights garnered from smart city development [33].

Urban Digital Twins (UDTs) are now being deployed globally, serving as valuable resources for understanding physical systems and spatial structures within smart cities, especially in terms of environmental sustainability and resilience [73]. These UDTs cover a wide array of systems such as power grids, renewable energy, transportation, water supply, and climatology. Moreover, the advancement of UDTs is intricately linked with other technologies like AI, the Internet of Things (IoT), Big Data, and edge computing, facilitating data collection and processing for better management and sustainability efforts [22].

Despite its recent rise, the concept of DT remains diverse and challenging to pin down due to its pluralistic nature. There is no one-size-fits-all definition or blueprint for DT, leading to a multitude of interpretations and focus across various domains like transport, governance, energy, construction, public safety, and public health [4, 17]. Ultimately, while UDTs strive to replicate physical systems in digital form, they are not exact replicas but rather models that aim to approximate reality [17]. This inherent variability and the quest for greater fidelity to real-world systems underscore the ongoing evolution and importance of DT and UDTs in shaping the future of urban environments [53].

The evolution of technologies and operations in the concept of digital twins for smart cities is illustrated in Fig. 2 Data collection in smart city digital twins involves leveraging technologies such as IoT (Internet of Things), cloud computing, and 5G networks. IoT devices collect data from various sources, including sensors, cameras, and other connected devices, which are then transmitted to the cloud for storage and processing. The high-speed, low-latency capabilities of 5G networks enable real-time data transmission and communication between devices [37], enhancing the efficiency of data collection.

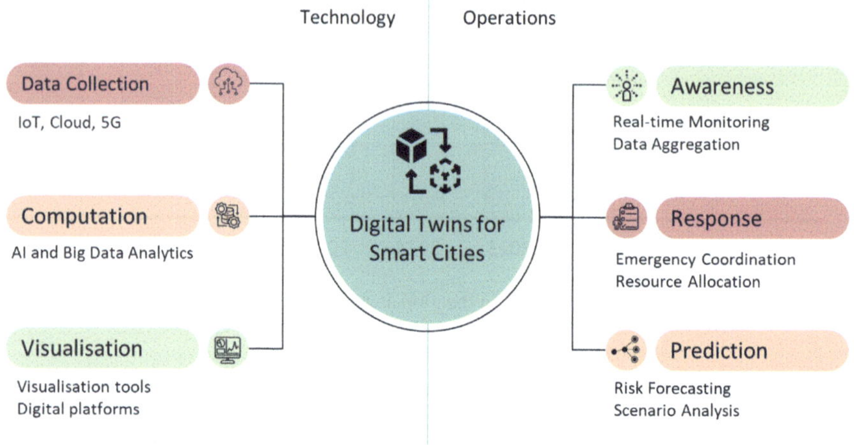

Fig. 2 Operations and technology enablers of city digital twins (*Source* inspired by IDC, 2020)

Computation plays a crucial role in smart city digital twins, with AI (Artificial Intelligence) and Big Data Analytics (BDA) being key components [7]. AI algorithms analyze vast amounts of data collected from IoT devices to extract meaningful insights and patterns, enabling data-driven decision-making [57]. BDA techniques further enhance this process by processing large datasets to uncover trends, correlations, and anomalies, providing valuable information for optimizing city operations and services [12].

Visualization is essential for understanding and interpreting the data collected and analyzed in smart city digital twins [16]. Visual representations, such as maps, 3D models, graphs, and dashboards, help stakeholders visualize complex data sets and trends, making it easier to identify patterns, anomalies, and potential areas for improvement [69]. Visualization tools provide intuitive interfaces for stakeholders to interact with and explore the data, enabling better decision-making and planning.

In terms of operations, smart city digital twins facilitate awareness, response, and prediction [40]. Awareness involves real-time monitoring and data aggregation to provide stakeholders with up-to-date information about various aspects of the city's operations and environment. Real-time monitoring systems collect and analyze data from IoT devices, sensors, and other sources to provide insights into traffic and mobility patterns, air quality, energy consumption, and more [67]. Data aggregation processes consolidate information from multiple sources to provide a comprehensive view of the city's status and performance [32].

Response capabilities enable emergency coordination and resource allocation in response to various events and situations [13]. Integrated systems use real-time data and analytics to identify emergencies, such as accidents, disasters, public safety, and security threats, and coordinate response efforts accordingly [26]. Resource allocation algorithms optimize the deployment of emergency services [55], personnel, and infrastructure to minimize response times and maximize effectiveness [17].

As mentioned, in recent years, AI has become essential in various decision-making processes such as planning, scheduling, machining, and inspection due to its domain independence and ability to enable self-adaptivity and self-learning within process planning systems [54]. Among emerging AI technologies, generative AI models like ChatGPT have gained attention (Fig. 3) for their capability to generate human-like text responses from extensive datasets [24]. In intelligent process planning, generative AI shares functionalities with knowledge-based systems, covering querying, interpreting, reasoning, and generating solutions [11]. However, while ChatGPT excels in providing information and proposing initial solutions, it may struggle with complex questions requiring critical analysis and intricate details within the smart city domain [62]. This underscores the need to develop dedicated generative AI models tailored for process planning, backed by rigorous validation processes to ensure the reliability of generated process knowledge or plans. Leveraging digital twin technology to validate process knowledge or plans generated by generative AI models can enhance their reliability and contribute to more effective decision-making processes [14].

GenAI significantly enhances the capabilities of Digital Twins by facilitating advanced simulations and predictive modeling [75]. GenAI algorithms can analyze extensive data from Digital Twins to identify patterns, trends, and potential outcomes [70]. Through machine learning and Deep Learning (DL) techniques, GenAI can produce descriptive realistic scenarios and simulations based on historical data, empowering Digital Twins to anticipate future events and optimize performance [46]. GenAI empowers the generation of virtual environments that mirror real-world conditions. This capability enables digital twins to carry out virtual experiments and simulations without the need for physical prototypes [42]. This empowerment enables Digital twins to make informed decisions, optimize processes, and enhance overall performance across domains such as manufacturing, healthcare, and urban planning [61].

Several research efforts have focused on designing and implementing digital twins to address the simulation and prediction of disasters in smart cities [9]. In these scenarios, GenAI has the potential to reduce costs and enhance efficiency [23].

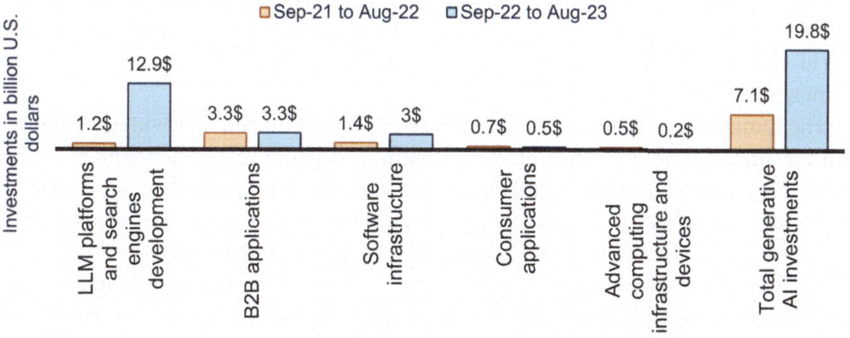

Fig. 3 U.S. investments in generative AI from 2021 to 2023 (*Source* Activate [1])

Empowering digital twins with GenAI in urban Disaster and Emergency Management (DEM) offers several benefits:

Enhanced Predictive Capabilities: GenAI algorithms can analyze vast amounts of data to predict potential disasters or their impacts with greater accuracy and foresight.

Realistic Scenario Simulation: GenAI facilitates the creation of realistic scenarios and simulations based on historical data, enabling digital twins to simulate various disaster scenarios and their potential outcomes.

Optimization of Response Strategies: Digital twins empowered with GenAI can optimize response strategies and resource allocation, enabling more efficient and effective emergency response efforts.

Rapid Decision-Making: Digital twins equipped with GenAI can provide real-time insights and recommendations, allowing decision-makers to quickly assess situations and make informed decisions during emergencies.

Improvement of Infrastructure Resilience: Through predictive modeling and scenario simulation, digital twins can identify vulnerabilities in urban infrastructure and systemic risks and suggest improvements to enhance resilience against disasters and their cascading impacts.

Cost Reduction: By simulating disaster scenarios in a virtual environment, digital twins can minimize the need for costly physical testing and experimentation, leading to cost savings in disaster preparedness and response efforts.

This chapter aims to bridge the gap between Generative Artificial Intelligence (GenAI) and digital twins within the context of DEM in smart cities. It aims to clarify how integrating GenAI enhances digital twins' predictive capabilities and simulation accuracy, especially in simulating disaster and emergency scenarios. chapter seeks to demonstrate its relevance and potential contributions to disaster management within urban environments through discussions on the principles, applications, and success stories of generative AI. Additionally, it aims to elucidate the evolution of digital twins and their pivotal role in predicting and mitigating disaster risks, supported by research works showcasing their effectiveness in disaster response. Furthermore, the chapter explores the intersection between generative AI and digital twins, explaining the functionalities that GenAI brings to the simulation process and providing real-world examples of successful integrations while addressing potential challenges and proposing solutions.

The remainder of this chapter is organized as follows: Sect. 2 provides a comprehensive literature review, thoroughly examining existing research pertinent to the topic. This section aims to establish a solid foundation by presenting an overview of the relevant scholarly works and analyzing which research gap this chapter fills. Section 3 explains the methodology employed in this study. Moving forward, Sect. 4 delves into the application of Digital twins in Smart City Disaster Management, exploring how these technologies are leveraged to enhance disaster preparedness and response operations. Following that, Sect. 5 focuses on Generative AI-enhanced digital twins, discussing how the integration of Generative Artificial Intelligence techniques enhances the capabilities of Digital twins for predictive modeling and optimization. Finally, Sect. 6 offers a concluding summary, highlighting the key findings and implications of the study while suggesting potential avenues for future research in the field of DTs within the smart city and DEM contexts.

2 Literature Review

2.1 Related Works

In recent years, transformative changes have been happening because of the fast growth of the telecommunication industry [8] and the widespread use of the IoT [31]. At the same time, huge advancements in data analytics, AI, and big data technologies have opened up new possibilities [2, 25, 58]. These enhancements have drawn attention to the concept of digital twins, leading to extensive research in the field [66]. In smart cities, where urban areas keep expanding rapidly, it becomes really important to make cities more resilient to different kinds of hazards. There are so many challenges, from earthquakes and volcanos [74] to wildfires [76], flooding and hurricanes [5], industrial and transportation accidents [50], and transportation and energy systems failures [36]. The COVID-19 pandemic [41] has shown how important it is to be ready for unexpected health emergencies in interconnected urban systems [19].

To tackle these challenges, many researchers have embraced developing digital twins. These models go beyond simply replicating cities digitally; they aim to grasp the intricacies of urban functionality and devise improved strategies for handling disasters as they arise. These models range from the scale of buildings to modeling subsystems within a city. In Hyun et al. [32], authors introduce a practical approach for integrating Building Information Modeling (BIM) and Geographic Information System (GIS) data to create a general-purpose digital twin for smart cities. The proposed methodology lays the foundation for diverse smart city services like disaster management and energy optimization, promising improved urban resilience and citizen well-being. Huang et al. [30] in another research contribute to the understanding and application of digital twin technology within integrated regional energy systems of smart cities. They introduce a basic concept of digital twins and outline the construction method and potential applications of an energy internet digital twin. The authors of Ramos et al. [52] address the critical issue of water scarcity exacerbated by water loss in distribution systems due to insufficient knowledge, suboptimal design, and poor maintenance. It proposes a smart water grid with a digital twin to enhance monitoring, management, and system efficiency. Also, Ammar et al. [6] published a research paper addressing the transformative potential of Digital twins in revolutionizing asset data management in transportation offering a roadmap for implementation in other agencies [6]. Simulating urban transportation systems is also investigated by Niaz et al. [47] who explore the integration of digital twins and Industrial IoT technology in transportation infrastructure to enhance safety and automation. It emphasizes the importance of real-time remote monitoring and training for future self-automation systems.

In numerous recent studies, researchers have concentrated on using digital twins for disaster management in urban areas. This is driven by the recognition that a key motivation for deploying digital twins is to simulate crises, aiming to prevent costs and mitigate risks effectively. Ford and Wolf [27] presents a conceptual model for smart cities with digital twins specifically tailored for disaster management, emphasizing the integration of sensing and simulation across diverse infrastructure systems within communities. It suggests that leveraging community disaster management can provide a solid foundation for digital twin development. In another research, a flood prediction methodology is developed, integrating synchronization analysis and deep learning to directly simulate the complex relationships between rainfall and flood characteristics [28]. Sihan Yu and coauthors introduced the concept of a digital twin for earthquake disaster cities, utilizing advanced technologies to simulate earthquake scenarios effectively. Real-time sensor monitoring enables dynamic analysis within the digital twin, enhancing emergency rescue planning and accurate evaluation of earthquake disasters impacts [74]. A live fire tracking platform as a digital twin has been developed [38] to monitor active reported fires across over twenty cities in the U.S., as well as predict their smoke paths and impacts on the air quality of regions within their range. These studies demonstrate how various digital twins enhance urban disaster modeling and management capabilities.

Following the COVID-19 pandemic, there has been a significant surge in research attention towards similar types of disasters [51]. The unprecedented global impact of the pandemic has underscored the critical importance of understanding and effectively managing various types of disasters, ranging from health crises to natural hazard disasters and beyond [10].

Pesantez et al. [49] developed a digital twin that integrates Advanced Metering Infrastructure data with a hydraulic model to investigate the impacts of changes in water demands during the COVID-19 pandemic on water distribution systems. The research assesses the effects of altered consumption patterns by evaluating water quality, water age, pressure distribution, and energy consumption. Additionally, the study models a water supply interruption event to analyze network resiliency under different demand conditions. The digital twin framework offers a means to understand emerging consumption patterns and their implications for water system operation, providing valuable insights for water utility managers to optimize infrastructure operations and plan for long-term impacts. Another important research was conducted by Chen et al. [15]. This study explores how AI-enabled digital twins, can enhance pandemic preparedness and response efforts for COVID-2019 in smart cities. The study constructs a well-named model, COVID-19 by utilizing digital twins' technology to map real-world epidemic situations into virtual spaces and integrating blockchain and deep learning algorithms. Through simulation, the COVID-DT model demonstrates high network security performance with low latency, ensuring efficient and accurate information interaction for pandemic prevention and control in smart cities.

The studies mentioned above highlight a portion of the research endeavors focused on employing digital twins for disaster management in urban areas. Advanced technologies play a pivotal role in augmenting the performance of digital twins by enabling more accurate simulations, real-time data integration, and enhanced predictive capabilities. These technologies encompass a wide range of innovations, including but not limited to AI, machine learning, BDA, IoT, cloud computing, augmented reality (AR), and virtual reality (VR). Integration of these cutting-edge technologies empowers digital twins to simulate better and analyze complex systems, anticipate potential issues, optimize operations, and facilitate informed decision-making in various domains such as urban planning, infrastructure management, manufacturing, healthcare, and more. For example, we refer to some recent studies: Zohdi [77] that integrate machine learning with fire spread models, using discrete element and topographical sub-models to predict fire propagation and secondary fires. Their framework supports real-time simulations for first responders via machine learning-enabled digital twins. In Yang et al. [72], the authors introduce digitally Twin Empowered IoT (DTEI), merging digital twins and IoT to advance Industry 4.0. They propose optimizing Federated Learning (FL) for DTEI by developing a DTEI-assisted deep reinforcement learning method and an asynchronous FL scheme, addressing data transmission and privacy concerns. Through federated learning, the facilitation of collaborative model training in urban environments, where data is sourced from diverse sensors and systems across different entities, is highlighted by Liu et al. [43]. Another research paper proposes a framework that integrates digital twins of cities with FL to address urgent crises such as the COVID-19 pandemic [48]. By leveraging FL, which allows collaborative learning while preserving data privacy, the framework enables multiple city DTs to share local strategies and statuses quickly. An FL central server manages updates from various city DTs, training a global model iteratively to correlate response plans with infection trends. The authors [53] contribute by exploring the integration of FL and DT technologies within smart city applications. Their research highlights the potential of FL to address privacy and trust concerns associated with sharing sensitive data in DT-based systems. Through an extensive survey of FL models in DT-based smart city applications, the paper identifies challenges and proposes future directions to enhance the integration of FL and DT. Overall, the study offers insights for effectively leveraging FL and DT technologies in real-time, life-critical scenarios, and for governance in smart city applications.

Li et al. [39] endeavor to enhance smart city governance and optimize data processing efficiency by harnessing BDA within the IoT framework. To tackle the complex multi-source data characteristic of smart cities, the research integrates DL algorithms with BDA and introduces a distributed parallelism strategy for convolutional neural networks. Furthermore, digital twins and multi-hop transmission technology are incorporated to establish a smart city DTs multi-hop transmission IoT-BDA system based on DL, with performance evaluation conducted through simulation. This research demonstrates how DL in smart city IoT-BDA systems reduces data delay, improves forecasting, and boosts practical efficiency, informing future smart city digital development.

Efficient data processing in smart cities, especially for emergency services, is addressed by Wolf et al. [68]. A prototype utilizing Microsoft Azure cloud computing technology is developed to provide analytical functionalities, allowing stakeholders to access real-time traffic and weather data for more effective emergency response coordination. The visualization function of digital twins is crucial as it enables stakeholders to comprehensively understand complex systems, identify patterns, and make informed decisions based on real-time data insights [20]. In Sharifi et al. [59], the authors explore the utility of digital twins as advanced virtual representations that offer deep insights into physical objects, particularly within smart city water systems. It emphasizes the importance of digital twins in various research and industrial sectors, highlighting their potential for improving data analysis and visualization techniques through machine learning.

2.2 Research Gap

With the emergence of GenAI, its potential in developing digital twins has been explored in numerous studies, including [14, 23, 42, 46, 70, 75]. The contribution of Bordukova et al. [14] lies in addressing the potential integration of GenAI techniques. The paper introduces a conceptual framework that synergistically incorporates digital twins. The framework is supported by two methodologies, including process generative pre-trained transformer modeling and a digital twin-based process verification method. By leveraging GenAI techniques within the framework of digital twins, the paper aims to enhance process planning capabilities and optimize manufacturing operations. This integration allows for the generation of synthetic data to supplement real-world observations, enabling more accurate simulations and predictive modeling. Despite efforts in AI-empowered digital twins, only a few studies have investigated the capabilities of Gen AI within digital twin frameworks for covering urban management issues including emergencies.

This gap highlights the need for exploration into Gen AI's potential for enhancing digital twins in disaster management within smart cities. While digital twins have shown promise in improving urban resilience and response strategies, the integration of GenAI techniques remains largely untapped. GenAI, with its advanced machine learning algorithms capable of generating novel and contextually relevant data, offers an opportunity to bolster digital twins' capabilities in simulating and predicting complex urban disasters. Leveraging GenAI, digital twins could dynamically adapt to evolving disaster scenarios, generate synthetic data to complement limited real-world observations and simulate diverse disaster scenarios to assist in proactive planning and response efforts. Moreover, GenAI's capacity to generate alternative scenarios and optimize decision-making processes could substantially enhance the effectiveness of disaster management strategies in smart cities. Consequently, further research is necessary to explore how GenAI can be effectively utilized to empower more resilient and adaptive digital twins for disaster management in smart urban environments. To fill the gap, this chapter delves into utilizing GenAI to enhance digital twins for disaster management in smart cities. It explores how GenAI improves simulation processes, addresses challenges, and suggests solutions.

3 Methodology

This study investigates the integration of GenAI with digital twins for disaster management in smart cities based on the methodology proposed by Klerings et al. [35]. Relevant insights are gathered after an extensive literature review across academic databases and journals encompassing digital twins, GenAI, disaster management, and smart cities. Through meticulous examination, existing knowledge gaps and areas necessitating further exploration in utilizing GenAI with digital twins for disaster management are identified. Subsequently, research questions are formulated, and specific examples are selected for in-depth analysis. The findings provide a solid foundation for future research endeavors in leveraging GenAI for enhanced disaster management through digital twins. The following steps explain the research methodology:

(1) **Identifying Relevant Literature**: Conduct a thorough search of academic databases, journals, conference proceedings, and reputable sources to gather literature on digital twins [33], Generative AI [14], disaster management [44], and smart cities [31]. Use keywords such as "digital twins," "Generative AI," "disaster management," "smart cities," "simulation," "predictive modeling," and "urban intelligence."
(2) **Reviewing Existing Studies**: Analyze existing literature to understand the current state-of-the-art in digital twin technology, Generative AI, and their applications in disaster management within smart cities. Identify key concepts, theories, methodologies, and findings relevant to the research topic.
(3) **Identifying Gaps and Challenges**: Identify gaps, limitations, and challenges in the existing literature related to the integration of Generative AI with digital twins for disaster management in smart cities. Determine areas where further research is needed and potential barriers to implementation.
(4) **Synthesizing Findings**: Synthesize findings from the literature review to develop a conceptual framework for understanding the synergies between Generative AI and digital twins in disaster management. Identify theoretical frameworks, models, and methodologies that can guide the research.
(5) **Formulating Research Questions**: Based on the gaps identified in the literature review, formulate research questions that address the study's specific objectives. These questions should guide the subsequent research methodology and data collection process.

(6) **Critical Analysis and Evaluation**: Critically analyze the strengths and weaknesses of existing studies in integrating Generative AI with digital twins for disaster management. Evaluate the studies' methodologies, data sources, and findings to assess their reliability and validity.
(7) **Identifying Emerging Trends**: Identify emerging trends, innovations, and future directions in the field of Generative AI-enhanced digital twins for disaster management in smart cities. Discuss how these trends may shape the future of research and practice in the field.
(8) **Summarizing and Documenting Findings**: Summarize the findings of the literature review coherently, highlighting key insights, trends, and gaps identified. Document the sources used in the review and provide a comprehensive bibliography for reference. enhanced digital twins in smart cities to handle disasters.

4 Digital Twins in Smart City Disaster Management

In the dynamic landscape of urban development and disaster management, the concept of digital twins has emerged as a transformative tool, particularly in the realm of smart cities [31]. In the context of smart city disaster management, digital twins offer a powerful platform for simulation, analysis, and decision-making [59]. They provide a comprehensive understanding of urban infrastructure, including buildings, transportation networks, utilities, and environmental conditions. By integrating data from various sources such as sensors, IoT devices, satellite imagery, and social media feeds (Fig. 4), digital twins enable city planners and emergency responders to anticipate, mitigate, and respond to disasters more effectively [74].

One key advantage of digital twins in disaster management is their ability to facilitate predictive modeling and scenario analysis. By simulating different disaster scenarios, authorities can assess vulnerabilities, identify critical points of failure, and devise strategies for resilience and recovery [32]. For instance, digital twins can

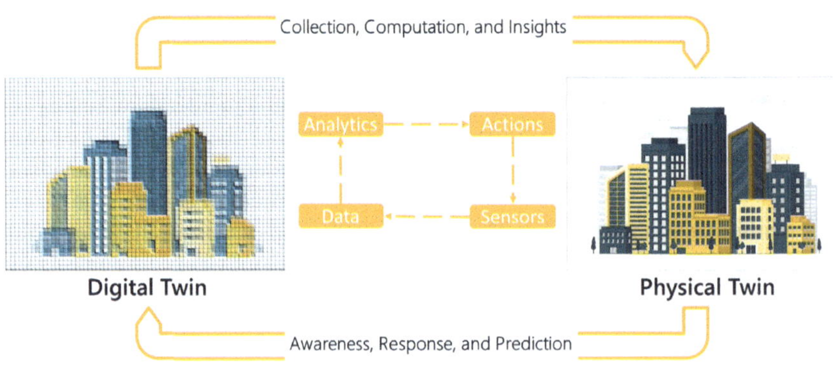

Fig. 4 Smart cities drive adaptable twins with data insights

simulate the impact of natural disasters like floods, earthquakes, or hurricanes on urban infrastructure [34], helping to optimize evacuation routes, allocate resources, and coordinate emergency response efforts [9].

Moreover, digital twins support real-time monitoring and situational awareness during disasters. By continuously updating with live data feeds, they provide a holistic view of the unfolding situation, enabling decision-makers to make informed choices in rapidly evolving circumstances [76]. For example, during a wildfire outbreak, digital twins can monitor fire spread, assess air quality, and track the movement of people and assets, allowing authorities to issue timely warnings and deploy resources where they are most needed [76]. Furthermore, digital twins facilitate post-disaster recovery and reconstruction efforts by enabling accurate damage assessment and resource allocation [27]. By comparing pre-event and post-event data, city planners can prioritize restoration efforts, streamline rebuilding processes, and ensure that critical infrastructure is restored in a resilient and sustainable manner [9].

Disasters can be categorized by response time from slow to very fast [74], and the relationship between functionality and the speed of disaster response is crucial in determining the effectiveness of digital twin applications in mitigating and managing urban crises [45]. Table 1 categorizes disaster response scenarios based on their speed, from very fast to slow, and outlines corresponding examples alongside digital twin functionalities.

In situations requiring very fast response, such as earthquakes [74] or flash floods, digital twins excel in providing real-time data insights and predictive capabilities. These functionalities enable rapid evacuation simulations and immediate flood prediction models, essential for minimizing casualties and damage [32] for slow-response challenges like climate change or population displacement, digital twins play a pivotal role in long-term planning and community engagement, laying the groundwork for sustainable resilience strategies. The adaptability and versatility of

Table 1 The functionalities of digital twins are categorized by the speed of disaster response

Disaster response speed	Example	Digital twin functionality
Very fast-response	Earthquakes, terrorist attacks, flash floods	Real-time sensor networks for earthquake early warning systems, rapid evacuation simulations, Immediate flood prediction models
Fast-response	Hurricanes, industrial accidents	Dynamic traffic management systems, real-time air quality monitoring, emergency communication platforms
Moderate-response	Wildfires, transportation accidents	Building damage assessment models, real-time monitoring and control systems for critical infrastructure, GIS-based decision support systems
Slow-response	Climate change, population displacement	Long-term urban planning and resilience strategies, community engagement platforms, urban redevelopment and reconstruction simulations

digital twin functionalities across varying speeds of disaster response underscore their importance in enhancing urban preparedness and resilience [44]. For instance, in moderate response scenarios such as hurricanes or pandemics, digital twins can support dynamic resource allocation, infrastructure resilience assessments, and coordinated response efforts [45]. By simulating the spread and impact of these events, digital twins enable authorities to optimize emergency operations and coordinate with multiple agencies efficiently [74]. Their ability to analyze large datasets, visualize complex scenarios, and provide actionable insights makes them invaluable for all stages of disaster management.

Overall, digital twins represent a paradigm shift in smart city disaster management, offering a powerful blend of data-driven insights, predictive capabilities, and real-time monitoring to enhance resilience, response, and recovery in the face of urban disasters. As cities continue to grapple with the growing complexity and frequency of natural and man-made hazards, digital twins are poised to play a pivotal role in building safer, more resilient urban environments.

Within the domain of software and system engineering, different types of digital twins address specific facets of systems, assets, and processes, each fulfilling distinct roles across various industries. Each type serves specific purposes, contributing to advancements and efficiencies across various domains [67]:

- **Component Twin**: Focuses on individual elements within a system, crucial for optimization and maintenance in industries like manufacturing.
- **Product/Asset Digital Twin**: Offers a holistic view of a product or asset's lifecycle, essential for reliability and efficiency in sectors such as aerospace and construction.
- **System Twin**: Captures interactions and dynamics among interconnected systems, vital for optimization in fields like urban planning and energy management [67].
- **Process Digital Twin**: Models and simulates specific processes or workflows, enabling analysis and optimization in industries like healthcare and manufacturing [45].

Table 2 presents an overview of different types of digital twins and their applications across various phases of disaster management. Each row corresponds to a specific phase, including Prevention/Mitigation, and Preparedness, Response, Recovery, and Reconstruction. The columns represent the four different types of digital twins.

Table 2 Digital twin applications across disaster management phases

DM phase	Type of digital twin			
	component twin	Product/asset DT	System twin	Process DT
Prevention	Analyze historical data for improvement	Analyze historical data for prevention	Identify systemic risks for prevention	Identify vulnerabilities for prevention
Mitigation	Simulate failure scenarios for planning	Simulate asset failures for mitigation	Simulate system failures for mitigation	Simulate process failures for mitigation
Preparedness	Provide training simulations for personnel	Train personnel on operations	Conduct drills for emergency preparedness	Provide training simulations for personnel
Response	Monitor component performance	Monitor asset performance	Optimize system performance	Optimize process workflows
Recovery	Monitor repaired components	Monitor repaired assets	Optimize system recovery	Optimize resource allocation
Reconstruction	Assist in rebuilding/ upgrading components	Assist in redesigning/ upgrading assets	Assist in redesigning systems	Assist in redesigning/ reengineering processes

5 Generative AI-Enhanced Digital Twins

5.1 Generative AI

GenAI encompasses technology that leverages deep learning models to generate content resembling human output, spanning images and text, in response to various prompts. A notable instance is ChatGPT, a GenAI-driven chatbot that has sparked discussions regarding its capacity to produce human-like text based on provided prompts or contextual cues. Researchers and developers can tailor these models (see Table 3) to suit particular domains and address specific tasks by refining the generative AI's learning algorithms or model architectures. The table provides a comparison of various Generative AI models, including their training data sources, notable features, and creators. Each model offers unique capabilities, from generating human-like text to creating photorealistic images from textual descriptions.

Generative AI models operate through a structured process, encompassing several key stages to effectively produce new content based on learned patterns from extensive datasets [11]:

Table 3 Varieties of popular generative AI models

Model	Training data	Notable features	Creator
GPT	Unsupervised pre-training on vast text data	Transformer-decoder layers, fine-tuning for specific tasks	OpenAI
GPT-2	Trained on various datasets beyond web text	Advanced results with zero-shot learning	OpenAI
GPT-3	Pre-trained on vast text data	Few-shot or zero-shot learning, the linguistic structure of text probability	OpenAI
GPT-4	Pre-trained on vast text data	Human-like performance, multimodal generation	OpenAI
LLaMA	Social media posts, web pages, books, news articles	Supports meta applications	Meta
Palm 2	Multilingual text and images	Multilingual support, zero-shot learning	Google
BLOOM	Diverse data sources including various languages	Proficiency in 46 natural languages, 13 programming languages	Bloom AI
BERT	Extensive textual data	Bidirectional self-attention, pre-trained model	Google
DALL-E 2	Not specified	Converts textual descriptions into images using a diffusion model	OpenAI

- **Objective Definition and Data Gathering**: The process begins with a clear definition of the model's objective, specifying the type of content it aims to generate. This objective guides the collection of a diverse dataset aligned with the desired output [14, 56].
- **Data Preprocessing**: Before training, the dataset undergoes preprocessing to remove noise and errors, ensuring high-quality input for the model.
- **Model Architecture Selection**: Choosing the right model architecture is critical. Options like Generative Adversarial Networks (GANs), Variational Autoencoders (VAEs), and Transformers offer distinct advantages. Selection depends on the project's goals and dataset characteristics.
- **Implementation**: The selected model architecture is translated into code, defining its structure and connections. Frameworks such as TensorFlow and PyTorch provide resources to simplify this step.
- **Training**: Training involves iteratively exposing the model to the dataset, and refining its parameters to minimize the gap between generated and desired outputs. This phase demands significant computational resources and time.
- **Evaluation and Optimization**: Post-training, the model's performance is evaluated using metrics to gauge content quality against expectations. If results fall short, adjustments to the architecture, training parameters, or dataset may be necessary for optimization [57].
- **Fine-Tuning and Iteration**: Continuous improvement is essential. Incorporating user feedback, introducing new data, and refining the training process through

iterations enhance the model's quality over time. Consistent refinement drives the development of a robust generative AI model.

5.2 Generative AI and Digital Twins

This section introduces the proposed architecture of the integration of Generative AI with Digital Twins. This architectural framework comprises three distinct layers, each serving a critical role in enhancing disaster management and urban planning capabilities. In the proposed architecture depicted in Fig. 5, the integration of Generative AI and Digital Twins is structured into three layers. At the edge layer, data is collected by sensors in urban areas, capturing parameters crucial for modeling the physical twin. Moving to the fog layer, this data undergoes processing, augmented by additional inputs such as topographical maps, historical records, climate data, and real-time social media updates. Within the fog layer, federated learning techniques enhance analysis by harnessing decentralized data sources, ensuring a collaborative model training process across distributed devices. Finally, the cloud layer employs AI and machine learning algorithms to analyze the integrated data, incorporating insights generated from federated learning in the fog layer. This cohesive integration enables the generation of real-time predictions with heightened accuracy and efficiency.

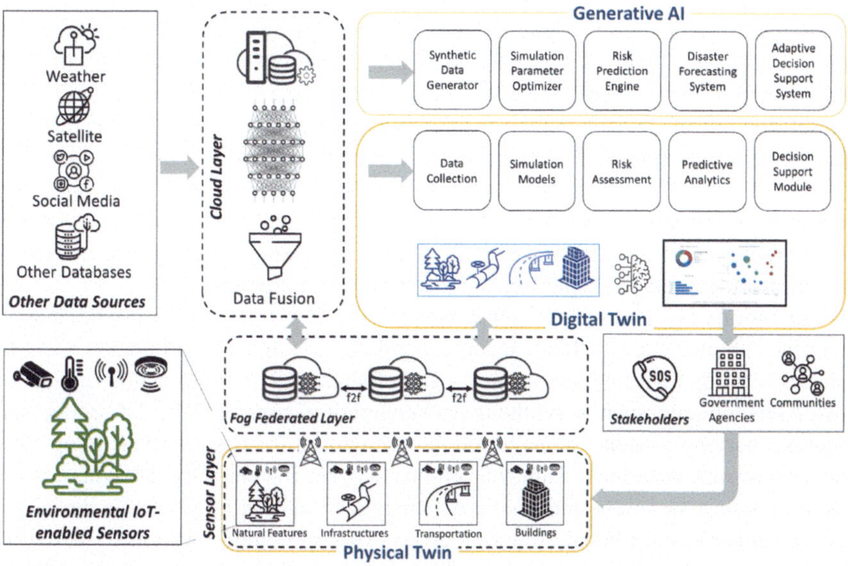

Fig. 5 GenAI-empowered digital twin architecture

Table 4 Corresponding components in DT and GenAI

Digital twin		Generative AI	
Data collection	Gathering real-time and historical data on urban infrastructure, population, and environmental conditions	Synthetic data generator	Generative adversarial networks (GANs) for creating realistic synthetic data
Simulation models	Building models to simulate various disaster scenarios such as earthquakes, floods, fires, etc	Simulation parameter optimizer	AI-driven optimization of simulation parameters for accuracy and speed
Risk assessment	Analyzing the vulnerability of urban infrastructure, population density, and critical facilities to different types of hazards	Risk prediction engine	Machine learning algorithms for predictive risk modeling and scenario analysis
Predictive analytics	Using data and models to predict the likelihood and impact of future disaster events	Disaster forecasting system	AI algorithms for advanced forecasting and trend analysis
Decision support	Providing tools and insights to help decision-makers plan and respond effectively to disasters	Adaptive decision support system	AI-based decision support systems for real-time adaptive planning and response strategies

As illustrated in Fig. 5, the integration of GenAI and DT reveals a dynamic interplay between corresponding components, each empowering the other to enhance functionality and efficacy. Table 4 serves as an exposition of these key components, elucidating their roles and the synergistic relationships between traditional and AI-driven approaches. From the foundational aspects of data collection to the pivotal domain of decision support, the table navigates critical facets of disaster management, illuminating the transformative potential inherent in the integration of AI technologies. This comparative analysis underscores the evolutionary trajectory of disaster management practices, wherein advancements in AI, such as synthetic data generation and machine learning algorithms, are reshaping predictive analytics, risk assessment methodologies, and decision-making paradigms.

Integration of GenAI with DT empowers Digital Twins in urban disaster management by imbuing them with enhanced capabilities to simulate and predict various disaster scenarios with greater accuracy and efficiency. The combination of Generative AI techniques, such as synthetic data generation, augments the data collection process, ensuring a comprehensive and diverse dataset for Digital Twins to operate upon. Moreover, machine learning algorithms integrated within Digital Twins enable real-time risk assessment and predictive analytics, facilitating proactive decision-making and resource allocation during crises. This symbiotic relationship between GenAI and DT fosters a dynamic feedback loop wherein insights generated from AI-driven analysis further refine the modeling and simulation capabilities of Digital

Twins, thereby continuously improving disaster response strategies. As a result, the integration of GenAI and DT emerges as a cornerstone in modern urban disaster management, offering a potent toolkit for resilience-building and effective mitigation of urban hazards. In Table 5, a comparative analysis is presented between AI-enabled Digital Twins and GenAI-Empowered Digital Twins across critical aspects relevant to urban disaster management. This comparison highlights the distinct functionalities and capabilities of each type of Digital Twin, shedding light on their respective strengths in data collection, decision-making processes, scalability, resource requirements, predictive capabilities, adaptability to uncertainty, and integration with other technologies. Through this comprehensive examination, valuable insights into the potential contributions of AI-driven technologies to enhance disaster management strategies are provided.

Table 5 Comparison between AI-enabled DT and GenAI-empowered DT

Aspects	AI-enabled DT	GenAI-empowered DT
Data collection	Relies on structured or manual data sources, with some incorporation of AI algorithms for data analysis	Leverages a wide range of data sources, including structured, unstructured, and real-time data streams, with AI algorithms automating data collection and integration processes
Decision-making process	Supports decision-making by providing insights or recommendations based on AI analysis of data	Utilizes AI algorithms for automated decision-making or to support human decision-makers with data-driven insights, predictions, and optimization
Scalability and flexibility	Offers limited scalability and flexibility, with some ability to adapt to changing conditions based on AI models	Provides greater scalability and flexibility through AI-driven automation and adaptation, allowing for more dynamic responses to changing conditions and larger-scale deployments
Cost and resource requirements	Requires investments in AI technologies and computational resources, but may be less resource-intensive compared to GenAI-Empowered DT	Requires significant investments in AI technologies and computational resources for data processing, modeling, and optimization, but offers greater potential for cost savings and efficiency gains in the long run
Accuracy and predictive capabilities	Provides predictive capabilities for specific use cases or scenarios based on AI models and historical data	Offers advanced predictive modeling and forecasting capabilities across multiple domains, leveraging AI to anticipate future events, trends, and outcomes with higher accuracy and precision
Adaptability to Uncertainty	May struggle to adapt to unforeseen circumstances or changing conditions due to limited learning capabilities	Is more adaptable and resilient due to continuous learning and improvement over time, enabling better handling of uncertainty and unexpected events

(continued)

Table 5 (continued)

Aspects	AI-enabled DT	GenAI-empowered DT
Integration with other technologies	Operates independently or requires manual integration with other systems and technologies	Seamlessly integrates with AI-driven analytics platforms, IoT devices, and other emerging technologies, enabling more comprehensive and efficient data-driven solutions
Explainability and reasoning	Limited explainability and reasoning capabilities, with decisions primarily based on data analysis without detailed explanations	Incorporates advanced methods for explaining AI-generated decisions and reasoning processes, enhancing transparency and understanding of decision-making outcomes

6 Conclusion

In conclusion, this chapter provides an overview of Generative Artificial Intelligence (GenAI) capabilities in empowering digital twins for disaster management in urban areas. This chapter explores the multifaceted potential of GenAI-enhanced digital twins in disaster management with an examination of the synergies between GenAI and digital twins, showcasing how these technologies can predict, mitigate, and respond to disasters in urban environments.

Also, the chapter presented an architecture for integrating GenAI with digital twins and compared different types of digital twins in disaster management. Additionally, the chapter compared traditional AI-enabled digital twins with GenAI-empowered digital twins. By examining the potential of GenAI in enhancing digital twins for disaster management, this chapter highlights the transformative impact of integrating generative algorithms into existing systems. The architecture presented offers insights into the seamless integration of GenAI with digital twins, facilitating more accurate predictions and simulations of disaster scenarios.

Furthermore, the comparison of different types of digital twins in disaster management underscores the unique advantages and applications of GenAI-empowered digital twins. By leveraging generative algorithms, these digital twins can generate synthetic data and simulate realistic disaster scenarios, enhancing urban resilience and response capabilities.

Moreover, the comparison between traditional AI-enabled digital twins and GenAI-empowered digital twins sheds light on the significant advancements offered by GenAI. While traditional AI-enabled digital twins rely on predefined rules and algorithms, GenAI-empowered digital twins can generate new data and adapt to changing conditions, making them more versatile and effective in disaster management scenarios.

Overall, this chapter emphasizes the potential of GenAI-enhanced digital twins to revolutionize disaster management practices in smart cities. By harnessing the power of generative algorithms, these digital twins can improve decision-making, enhance resilience, and ultimately save lives in the face of adversity. Also, this work identifies a research gap in the literature and suggests several future research directions. These include exploring dynamic integration architectures, studying the architecture of integrating GenAI with digital twins, conducting focused studies on specific GenAI-empowered digital twins designed for pre and post-disaster actions, investigating the ethical implications of employing GenAI in disaster management, examining the scalability and performance of GenAI-enhanced systems in real-world scenarios, and exploring novel applications of GenAI in interdisciplinary domains such as healthcare and environmental monitoring. These avenues for further research are crucial for advancing understanding and implementation in the field of disaster management and extending the applicability of GenAI across diverse domains.

References

1. Activate (2023) Investments in generative artificial intelligence (AI) in the United States from September 2021 to August 2023, by category (in billion U.S. dollars) [Graph]. In: Statista. Retrieved April 16, 2024, from https://0-www-statista-com.biblioteca-ils.tec.mx/statistics/1454260/united-states-generative-ai-investments-by-category/
2. Ajulo O, Asgary A, Tang P, Von-Meding J (2021) Modeling transformative adaptation: case of post-earthquake Lyttelton, New Zealand. Environ Sci Policy 125:247–262
3. Al Mtawa Y, Haque A, Bitar B (2019) The mammoth internet: are we ready? IEEE Access 7:132894–132908
4. Al-Sehrawy R, Kumar B, Watson R (2023) The pluralism of digital twins for urban management: bridging theory and practice. J Urban Manage 12(1):16–32
5. Amiri A, Soltani K, Ebtehaj I, Bonakdari H (2024) A novel machine learning tool for current and future flood susceptibility mapping by integrating remote sensing and geographic information systems. J Hydrol 130936
6. Ammar A, Maier F, Pratt W S, Richard E, Dadi G (2024) Practical application of digital twins for transportation asset data management: case example of a safety hardware asset. Transp Res Rec, 03611981241231804
7. Ardakani SP, Cheshmehzangi A (2024) Big data analytics for smart transport and healthcare systems. Springer
8. Ardakani SP, Kwong CF, Kar P, Liu Q, Li L (2021) CNN: a cluster-based named data routing for vehicular networks. IEEE Access 9:159036–159047
9. Ariyachandra MMF, Wedawatta G (2023) Digital twin smart cities for disaster risk management: a review of evolving concepts. Sustainability 15(15):11910
10. Asgary A, Aarabi M, Dixit S, Wen H, Ahmed M, Wu J (2024) A survey of the use of modeling, simulation, visualization, and mapping in public health emergency operations centers during the covid-19 pandemic. Int J Environ Res Public Health 21(3):295
11. Barcaui A, Monat A (2023) Who is better at project planning? generative artificial intelligence or project managers? Proj LeadShip Soc 4:100101
12. Bi Z, Jin Y, Maropoulos P, Zhang WJ, Wang L (2023) Internet of things (IoT) and big data analytics (BDA) for digital manufacturing (DM). Int J Prod Res 61(12):4004–4021
13. Bonadonna C, Frischknecht C, Menoni S, Romerio F, Gregg CE, Rosi M, Cristiani C (2021) Integrating hazard, exposure, vulnerability and resilience for risk and emergency management in a volcanic context: the ADVISE model. J Appl Volcanol 10:1–34
14. Bordukova M, Makarov N, Rodriguez-Esteban R, Schmich F, Menden MP (2024) Generative artificial intelligence empowers digital twins in drug discovery and clinical trials. Expert Opin Drug Discov 19(1):33–42

15. Chen D, AlNajem NA, Shorfuzzaman M (2022) Digital twins to fight against COVID-19 pandemic. Internet Things Cyber-Phys Syst 2:70–81
16. Chen K, Nadirsha TN, Lilith N, Alam S, Svensson Å (2024) Tangible digital twin with shared visualization for collaborative air traffic management operations. Transp Res Part C: Emerg Technol 161:104546
17. Cheng R, Hou L, Xu S (2023) A review of digital twin applications in civil and infrastructure emergency management. Buildings 13(5):1143
18. Cheshmehzangi A, You N, Siri J, Birch E (2024) The nexus between urban innovation and sustainable development goals (SDGs). In: Harnessing urban innovation to unlock the sustainable development goals. Springer Nature Singapore, Singapore, pp 1–15
19. Champlin C, Sirenko M, Comes T (2023) Measuring social resilience in cities: an exploratory spatio-temporal analysis of activity routines in urban spaces during Covid-19. Cities 135:104220
20. Comba JL, Santos NO, Rivera JC, Romeu RK, Abel M (2023) Data visualization for digital twins. Comput Sci Eng 25(2):58–63
21. Cui HY, Cao YQ (2024) Do smart cities improve energy efficiency? a test of spatial effects and mechanisms. Sustain Cities Soc 101:105124
22. Dihan MS, Akash AI, Tasneem Z, Das P, Das SK, Islam MR., Hasan MM (2024) Digital twin: data exploration, architecture, implementation and future. Heliyon
23. Du B, Du H, Liu H, Niyato D, Xin P, Yu J., Tang Y (2023) YOLO-based semantic communication with generative AI-aided resource allocation for digital twins construction. IEEE Internet Things J
24. Dwivedi YK, Pandey N, Currie W, Micu A (2024) Leveraging ChatGPT and other generative artificial intelligence (AI)-based applications in the hospitality and tourism industry: practices, challenges and research agenda. Int J Contemp Hosp Manag 36(1):1–12
25. Ebtehaj I, Bonakdari H (2024) Generalized structure of group method of data handling: Novel technique for flash flood forecasting. Water Resour Manage 1–19
26. Fan C, Zhang C, Yahja A, Mostafavi A (2021) Disaster city digital twin: a vision for integrating artificial and human intelligence for disaster management. Int J Inf Manage 56:102049
27. Ford DN, Wolf CM (2020) Smart cities with digital twin systems for disaster management. J Manage Eng 36(4):04020027
28. Ghaith M, Yosri A, El-Dakhakhni W (2022) Synchronization-enhanced deep learning early flood risk predictions: the core of data-driven city digital twins for climate resilience planning. Water 14(22):3619
29. Grieves MW (2023) Digital twins: past, present, and future. The digital twin. Springer International Publishing, Cham, pp 97–121
30. Huang W, Zhang Y, Zeng W (2022) Development and application of digital twin technology for integrated regional energy systems in smart cities. Sustain Comput: Inform Syst 36:100781
31. Huda NU, Ahmed I, Adnan M, Ali M, Naeem F (2024) Experts and intelligent systems for smart homes' transformation to sustainable smart cities: a comprehensive review. Expert Syst Appl 238:122380
32. Hyun C T, Lee S, Jin C (2024) Digital twin for disaster management in smart city. In: Future of information and communication conference. Springer Nature Switzerland, Cham, pp 627–641
33. Jafari M, Kavousi-Fard A, Chen T, Karimi M (2023) A review on digital twin technology in smart grid, transportation system and smart city: Challenges and future. IEEE Access 11:17471–17484
34. Kim J, Lee J M, Kang J (2023) Smart cities and disaster risk reduction in South Korea by 2022: the case of Daegu. Heliyon
35. Klerings I, Robalino S, Booth A, Escobar-Liquitay CM, Sommer I, Gartlehner G, Waffenschmidt S (2023) Rapid reviews methods series: guidance on literature search. BMJ Evid-Based Med 28(6):412–417
36. Kušić K, Schumann R, Ivanjko E (2023) A digital twin in transportation: real-time synergy of traffic data streams and simulation for virtualizing motorway dynamics. Adv Eng Inform 55:101858

37. Lessi CC, Gavrielides A, Solina V, Qiu R, Nicoletti L, Li D (2024) 5G and beyond 5g technologies enabling industry 5.0: network applications for robotics. Procedia Comput Sci 232:675–687
38. Lewis RH, Jiao J, Seong K, Farahi A, Navrátil P, Casebeer N, Niyogi D (2024) Fire and smoke digital twin—a computational framework for modeling fire incident outcomes. Comput Environ Urban Syst 110:102093
39. Li X, Liu H, Wang W, Zheng Y, Lv H, Lv Z (2022) Big data analysis of the internet of things in the digital twins of smart city based on deep learning. Futur Gener Comput Syst 128:167–177
40. Liao X, Zhao X, Wang Z, Zhao Z, Han K, Gupta R, Wu G (2023) Driver digital twin for online prediction of personalized lane change behavior. IEEE Internet Things J
41. Lieberman B, Kong JD, Gusinow R, Asgary A, Bragazzi NL, Choma J, Mellado B (2023) Big data-and artificial intelligence-based hot-spot analysis of COVID-19: Gauteng, South Africa, as a case study. BMC Med Inform Decis Mak 23(1):19
42. Liu J, Zhou Y, Li Y, Li Y, Hong S, Li Q, Wang X (2023) Exploring the integration of digital twin and generative AI in agriculture. In: 2023 15th international conference on intelligent human-machine systems and cybernetics (IHMSC). IEEE, pp 223–228
43. Liu K, Yan Z, Liang X, Kantola R, Hu C (2022) A survey on blockchain-enabled federated learning and its prospects with digital twin. Digit Commun Netw
44. Mitra A, Shaw R (2023) Systemic risk from a disaster management perspective: a review of current research. Environ Sci Policy 140:122–133
45. Mohammadi N, Taylor JE (2021) Thinking fast and slow in disaster decision-making with smart city digital twins. Nat Comput Sci 1(12):771–773
46. Mu H, He F, Yuan L, Hatamian H, Commins P, Pan Z (2024) Online distortion simulation using generative machine learning models: a step toward digital twin of metallic additive manufacturing. J Ind Inf Integr 38:100563
47. Niaz A, Khan S, Niaz F, Shoukat MU, Niaz I, Yanbing J (2022) Smart city IoT application for road infrastructure safety and monitoring by using digital twin. In: 2022 international conference on IT and industrial technologies (ICIT). IEEE, pp 1–6
48. Pang J, Huang Y, Xie Z, Li J, Cai Z (2021) Collaborative city digital twin for the COVID-19 pandemic: a federated learning solution. Tsinghua Sci Technol 26(5):759–771
49. Pesantez JE, Alghamdi F, Sabu S, Mahinthakumar G, Berglund EZ (2022) Using a digital twin to explore water infrastructure impacts during the COVID-19 pandemic. Sustain Cities Soc 77:103520
50. Pourroostaei Ardakani S, Cheshmehzangi A (2023) A predictive data analysis for traffic accidents: real-time data use for mobility improvement and accident reduction. In: Big data analytics for smart transport and healthcare systems. Springer Nature Singapore, Singapore, pp 81–99
51. Pourroostaei Ardakani S, Xia T, Cheshmehzangi A, Zhang Z (2022) An urban-level prediction of lockdown measures impact on the prevalence of the COVID-19 pandemic. Genus 78(1):28
52. Ramos HM, Kuriqi A, Besharat M, Creaco E, Tasca E, Coronado-Hernández OE, Iglesias-Rey P (2023) Smart water grids and digital twin for the management of system efficiency in water distribution networks. Water 15(6):1129
53. Ramu SP, Boopalan P, Pham QV, Maddikunta PKR, Huynh-The T, Alazab M, Gadekallu TR (2022) Federated learning enabled digital twins for smart cities: concepts, recent advances, and future directions. Sustain Cities Soc 79:103663
54. Rathore MM, Shah SA, Shukla D, Bentafat E, Bakiras S (2021) The role of ai, machine learning, and big data in digital twinning: a systematic literature review, challenges, and opportunities. IEEE Access 9:32030–32052
55. Razavi H, Ghasemi A (2017) Multi-Objective optimization of a qos-aware channel allocation in cognitive radio networks using NSGA-II algorithm. J Control 11(1):1–9
56. Razavi H, Jamali M R, Emsaki M, Ahmadi A, Hajiaghei-Keshteli M (2023) Quantifying the financial impact of cyber security attacks on banks: a big data analytics approach. In: 2023 IEEE Canadian conference on electrical and computer engineering (CCECE). IEEE, pp 533–538

57. Razavi H, Jamali MR, Emsaki M, Gholian-Jouybari F, Bonakdari H, Hajiaghaei-Keshteli M (2023) Statistical and data analytics approaches to parameter tuning for enhancing qos of e-banking transactions: a case study of sample bank. In: 2023 IEEE Canadian conference on electrical and computer engineering (CCECE). IEEE, pp 516–521
58. Salehi S, Kavgic M, Bonakdari H, Begnoche L (2024) Comparative study of univariate and multivariate strategy for short-term forecasting of heat demand density: exploring single and hybrid deep learning models. Energy AI 16:100343
59. Sharifi A, Beris AT, Javidi AS, Nouri MS, Lonbar AG, Ahmadi M (2024) Application of artificial intelligence in digital twin models for stormwater infrastructure systems in smart cities. Adv Eng Inform 61:102485
60. Statista (2023) Projected revenue generated by companies in the global smart city from 2020 to 2028 (in billion U.S. dollars) [Graph]. In: Statista. Retrieved March 29, 2024, from https://0-www-statista-com.biblioteca-ils.tec.mx/statistics/1111626/worldwide-smart-city-market-revenue
61. Sun Y, Zhang Q, Bao J, Lu Y, Liu S (2024) Empowering digital twins with large language models for global temporal feature learning. J Manuf Syst 74:83–99
62. Suri G, Slater L R, Ziaee A, Nguyen M (2024) Do large language models show decision heuristics similar to humans? a case study using GPT-3.5. J Exp Psychol General
63. Tao F, Zhang H, Zhang C (2024) Advancements and challenges of digital twins in industry. Nat Comput Sci 4(3):169–177
64. Ullah A, Anwar SM, Li J, Nadeem L, Mahmood T, Rehman A, Saba T (2024) Smart cities: the role of Internet of Things and machine learning in realizing a data-centric smart environment. Complex Intell Syst 10(1):1607–1637
65. Verma A, Verma V (2021) Comparative study of cloud computing and edge computing: Three level architecture models and security challenges. Int J Distrib Cloud Comput 9:13–17
66. Wang H, Wang Y (2024) Smart cities net zero planning considering renewable energy landscape design in digital twin. Sustain Energy Technol Assess 63:103629
67. Wang J, Li X, Wang P, Liu Q (2024) Bibliometric analysis of digital twin literature: a review of influencing factors and conceptual structure. Technol Anal Strat Manag 36(1):166–180
68. Wolf K, Dawson RJ, Mills JP, Blythe P, Morley J (2022) Towards a digital twin for supporting multi-agency incident management in a smart city. Sci Rep 12(1):16221
69. Wu J, Zhu J, Zhang J, Dang P, Li W, Guo Y, Liang C (2023) A dynamic holographic modelling method of digital twin scenes for bridge construction. Int J Digit Earth 16(1):2404–2425
70. Xu Q, Zhou G, Zhang C, Chang F, Cao Y, Zhao D (2023) Generative AI and digital twin integrated intelligent process planning: a conceptual framework
71. Yan Y, Kunhui Y (2024) Novel cyber-physical architecture for optimal operation of renewable-based smart city considering false data injection attacks: digital twin technologies for smart city infrastructure management. Sustain Energy Technol Assess 65:103733
72. Yang W, Xiang W, Yang Y, Cheng P (2022) Optimizing federated learning with deep reinforcement learning for digital twin empowered industrial IoT. IEEE Trans Industr Inf 19(2):1884–1893
73. Ye X, Du J, Han Y, Newman G, Retchless D, Zou L, Cai Z (2023) Developing human-centered urban digital twins for community infrastructure resilience: a research agenda. J Plan Lit 38(2):187–199
74. Yu S, Lei Q, Liu C, Zhang N, Shan S, Zeng X (2023) Application research on digital twins of urban earthquake disasters. Geomat Nat Hazards Risk 14(1):2278274
75. Zhang Y, He R, Ai B, Yang M, Chen R, Wang C, Zhong Z (2023) Generative adversarial networks based digital twin channel modeling for intelligent communication networks. China Commun 20(8):32–43
76. Zhong C, Cheng S, Kasoar M, Arcucci R (2023) Reduced-order digital twin and latent data assimilation for global wildfire prediction. Nat Hazard 23(5):1755–1768
77. Zohdi TI (2020) A machine-learning framework for rapid adaptive digital-twin based fire-propagation simulation in complex environments. Comput Methods Appl Mech Eng 363:112907

Optimizing Urban Heating: Integrating Geothermal Energy and Chemical Heat Pumps for Digital Twin Simulations

Zhengguang Liu, Masoud Babaei, Chen Chen Song, and Chenmeng Zhang

Abstract Chemical heat pump systems offer a promising avenue for harnessing geothermal energy and leveraging chemical reactions to release heat, thereby providing a viable, sustainable energy solution. Central to their success are thermal fluids, whose heat transfer properties and storage capabilities directly influence system performance. This chapter provides the first comprehensive evaluation of geothermal-chemical heat pump systems enhanced by Digital Twin technology, offering a deep understanding of operational mechanisms and the application of these systems. The systematic categorization of these systems includes both individual components and their integrated applications within a Digital Twin framework, highlighting the real-time data analysis and simulation capabilities that drive system optimizations. The significance of this chapter lies in its detailed meta-analysis, which reveals a surging interest in chemical heat pumps within geothermal energy, as evidenced by 409 publications. Despite this, there remains a need for more detailed research in numerous classifications. Fluorocarbons maintain their position with the highest COP values exceeding 4, while water stands out as the most cost-effective option, with a mere 0.12 USD per MJ and minimal environmental impact among thermal working fluids. The growing popularity of organic chemical heat pumps and nanofluids underscores the evolving landscape of thermal fluid choices. Discussions on selection criteria and performance ramifications of thermal fluids have been provided, focusing on diverse optimization methodologies aimed at bolstering system efficiency. By incorporating Digital Twin technology, this chapter identifies prevailing challenges and envisions prospects, thereby empowering the current landscape and providing a roadmap for further research and practical applications.

Z. Liu · M. Babaei
Department of Chemical Engineering, The University of Manchester, Manchester M13 9PL, UK

C. C. Song (✉)
Higher Information Industry Technology Research Institute, Beijing Information Science and Technology University, Beijing 100192, China
e-mail: songchenchen@bistu.edu.cn

C. Zhang
Institute of Population Research, Peking University, Beijing 100871, China

S. Pourroostaei Ardakani and A. Cheshmehzangi (eds.), *Digital Twin Computing for Urban Intelligence*, Urban Sustainability,
https://doi.org/10.1007/978-981-97-8483-7_6

Keywords Chemical heat pump · Geothermal energy · Renewable energy · Heat storage · Heat fluid

1 Introduction

Climate change presents an unparalleled threat to the stability and sustainability of global energy systems, urging a shift towards innovative and efficient approaches for energy conversion [1]. In response to this imperative, the collaborative integration of geothermal and chemical heat pump technologies, enhanced by Digital Twin technology, emerges as a beacon of promise in the quest for sustainable energy solutions. Geothermal energy, as a steadfast and renewable resource, draws its power from the Earth's inherent heat reservoirs [2]. This energy source holds immense potential, yet fully unlocking its capabilities demands the deployment of advanced technologies including Digital Twins. Geothermal energy's consistency stems from the Earth's natural heat flux, making it an ideal candidate for continuous and renewable power generation [3]. Harnessing this resource efficiently, however, requires addressing inherent challenges such as the variable nature of geothermal heat and the need for improved extraction methods, for which Digital Twin technology provides real-time simulation and predictive analytics. Installed geothermal power plant in 2022 (MW) has been shown in Fig. 1.

This is where the strategic synergy with chemical heat pumps, guided by Digital Twin insights, comes into play, offering a complementary partnership to optimize the utilization of geothermal energy [4]. These pumps leverage either chemical reactions

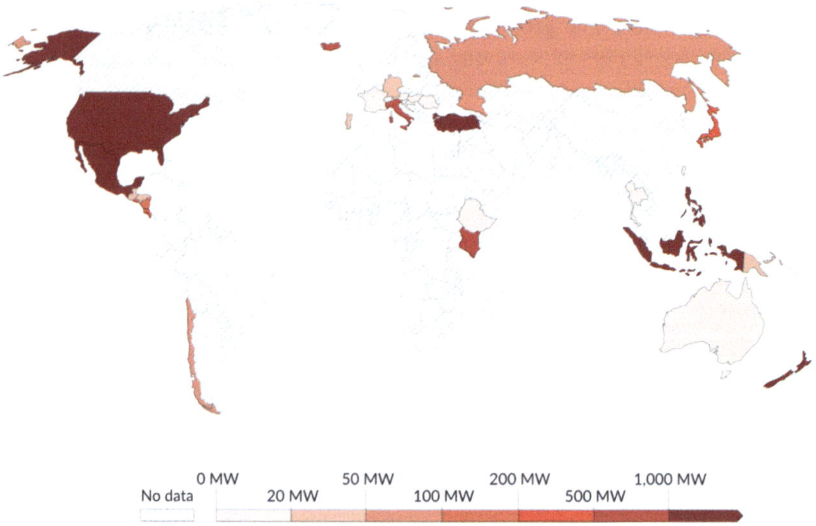

Fig. 1 Installed geothermal power plant in 2022 (MW)

or adsorption processes to amplify the efficiency of heat transfer and energy conversion, thereby mitigating challenges associated with the utilization of geothermal energy [5]. Chemical heat pumps, through their innovative mechanisms, serve as catalysts for refining the intricate dance between geothermal energy and the technologies required for its harnessing. By leveraging chemical reactions, these pumps facilitate a more controlled and intensified exchange of thermal energy [6, 7], monitored and optimized through Digital Twin technology. Alternatively, through adsorption processes, they adeptly capture and release heat, contributing to a more seamless and efficient utilization of the Earth's natural warmth.

Chemical heat pumps bring a unique set of advantages to the collaboration [8], utilizing either chemical reactions or adsorption processes to significantly enhance the efficiency of both heat transfer and energy conversion processes [9], with Digital Twins providing dynamic adjustments based on continuous data analysis. They enable the absorption and release of heat during chemical transformations, enhancing the overall efficiency of the geothermal system [10]. Through these mechanisms, chemical heat pumps not only augment the energy output from geothermal sources but also contribute to load balancing and grid stability, with the Digital Twin offering predictive insights that ensure continuous system performance. The amplification of heat transfer efficiency is a key facet of this collaboration [11], with the Digital Twin modeling variations in geothermal energy output and suggesting optimal operational strategies.

Revealing the fundamental role of heat fluid filtration and optimizing thermal energy storage, managed through the Digital Twin platform, emerges as a cornerstone in this collaborative endeavor [12]. The seamless integration of geothermal and chemical heat pump technologies relies on the meticulous orchestration of these two critical components, emphasizing their pivotal contributions to the overall synergy [13]. Heat fluid filtration plays a central role in ensuring the smooth functioning of the collaborative system [14]. It involves the meticulous purification and conditioning of the heat transfer medium, enhancing its efficiency in capturing and transporting thermal energy [15], with adjustments informed by the Digital Twin's predictive maintenance capabilities. Furthermore, the optimization of thermal energy storage constitutes a key pillar in fortifying the resilience and effectiveness of this cooperative framework [16], with strategic storage decisions driven by data insights from the Digital Twin.

Although there are individual studies on geothermal and chemical heat pumps, a comprehensive study on system integration, enhanced by Digital Twin technology, is still lacking [17]. This chapter aims to fill this gap by delving into the complementary nature of geothermal and chemical heat pumps in collaborative energy conversion and how optimizing heat fluid filtration, managed through Digital Twin simulations, can achieve more efficient overall energy conversion and heat storage [18]. This chapter focuses on the synergistic energy conversion of geothermal and chemical heat pump technologies, as well as the optimization of heat fluid filtration that can bring for thermal energy storage [19, 20], all under the oversight of a comprehensive Digital Twin framework. Previous literature has not thoroughly analyzed the specific mechanisms of heat fluid filtration throughout the entire heat conversion process [21]. This chapter will focus on in-depth research into the details of heat fluid filtration, including its applications in heat transfer processes and chemical reactions, addressing this knowledge gap, with the Digital Twin providing a continuously updated model that reflects current conditions and suggests optimizations. The objectives of writing this chapter are as follows:

(1) Consolidate fragmented knowledge on geothermal, chemical heat pumps, heat fluid filtration, and thermal energy storage, firstly providing a comprehensive and in-depth perspective through the lens of Digital Twin technology.
(2) By analyzing the current state of research in detail, highlight the strengths and weaknesses of existing studies, offering directions and insights for future research and steering research toward new directions, with simulations and predictions provided by Digital Twin models.
(3) Contribute to a deeper understanding of heat fluid filtration for thermal energy storage optimization, provides innovative and feasible solutions for the future transition to sustainable energy, facilitated by the advanced modeling and real-time data analysis capabilities of Digital Twin technology.

The fusion of geothermal energy and chemical heat pumps through digital twin technology represents a harmonious union of nature and technology. This strategic collaboration uses real-time data analysis and predictive modeling to optimize the use of sustainable energy and advance our efforts to achieve a sustainable and resilient energy landscape. This approach not only improves system efficiency, but also reduces dependence on fossil fuels and contributes to global efforts to reduce emissions.

Digital twin technology plays a key role in this process, dynamically adjusting the system through continuously updated data, managing the variability of geothermal energy, and enhancing the performance of chemical heat pumps. By revealing the core role of these components and their interactions, we gain a deeper understanding of the complex relationship between nature and technology, advancing our progress towards a sustainable and resilient energy system.

2 Geothermal-Chemical Heat Pump System

2.1 Geothermal Energy Supply Systems

Geothermal heating mechanisms, enhanced by Digital Twin technology, delve into the natural processes that lead to the extraction of heat from the Earth's interior. The Earth's subsurface harbors a significant amount of thermal energy generated by radioactive decay and residual heat from the planet's formation [22]. Geothermal power generation, modeled through Digital Twins, relies on the natural heat emanating from the Earth's interior. Heat is conducted from the Earth's depths to the surface layers through mechanisms such as heat conduction. Additionally, convective heat transfer in subsurface fluids, including magma and water, plays a crucial role, dynamically simulated within the Digital Twin to optimize energy extraction. Permeable rock formations further enhance the process by facilitating the upward movement of heated fluids [23], with Digital Twin models providing real-time insights into fluid dynamics and thermal gradients [24].

Geothermal power generation encompasses various technologies, including flash steam power plants, dry steam power plants, and enhanced geothermal systems, all visualized and analyzed through Digital Twin technology for optimal deployment in their potential distribution areas shown in Fig. 2. Dry steam power plants extract steam directly from geothermal reservoirs and employ it to drive turbines, generating electricity. This approach, often modeled in Digital Twins, boasts high efficiency and minimal environmental impact as it utilizes the Earth's natural steam. However, its application is limited to areas with substantial dry steam resources [26]. Flash steam power plants utilize high-pressure hot water from geothermal reservoirs to produce steam, which in turn drives turbines for electricity generation. This method is versatile and suitable for regions with diverse geothermal resources [27], with Digital Twin simulations helping to mitigate concerns regarding non-condensable gases and scaling issues. Binary cycle power plants use lower-temperature geothermal fluids to

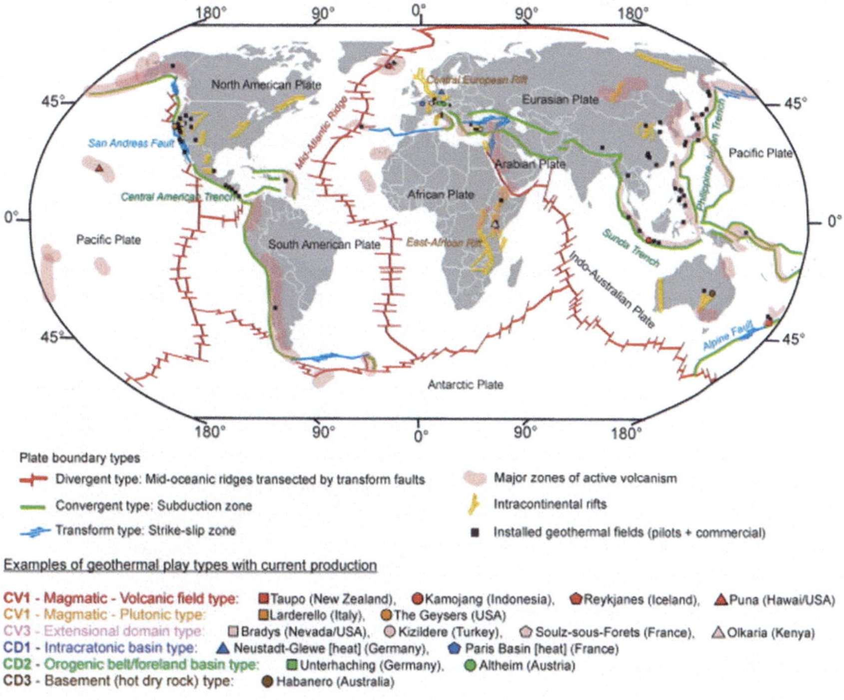

Fig. 2 Global Distribution of Geothermal Fields in Plates (CV: Convection, CD: conduction; *source* [25])

vaporize a secondary fluid with a lower boiling point, which then drives a turbine for electricity generation. This method, enhanced by Digital Twin predictions, allows for power generation in areas with moderate geothermal temperatures, expanding the geographical locations suitable for exploitation [28] but may exhibit lower efficiency compared to dry or flash steam systems. Enhanced Geothermal Systems (EGS) are designed to create or improve geothermal reservoirs by enhancing permeability through various technologies. Digital Twins aid in optimizing these enhancements, though challenges include technological complexity, potential induced seismicity, and high upfront costs.

The three main geothermal power generation technologies—Dry Steam, Flash Steam, and Binary Cycle—each have distinct advantages and disadvantages. Digital Twin technology plays a critical role in evaluating these technologies, offering insights into the optimal use of extracted steam for turbine operation, adapting to the presence of dissolved minerals in underground water, and minimizing environmental impacts using environmentally friendly organic substances. Ultimately, the choice among these technologies, as modeled and analyzed through Digital Twins, depends on the characteristics of underground heat resources, geographical conditions, economic factors, and environmental considerations. Figure 3 shows these technologies along with their advantages and disadvantages, analyzed through Digital Twin simulations to ensure the most efficient and sustainable deployment.

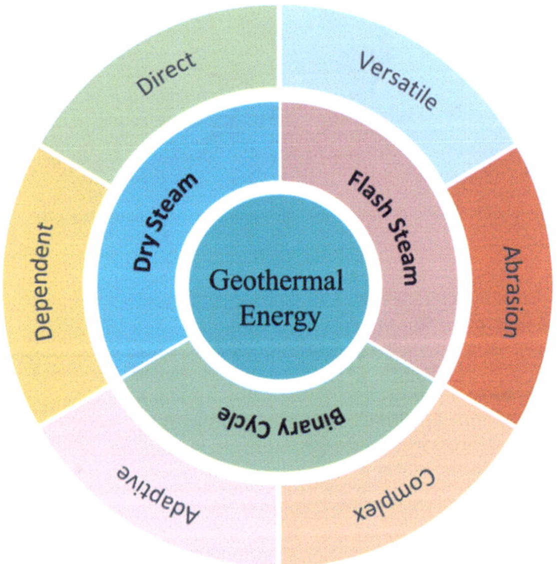

Fig. 3 Three main geothermal power generation technologies and their advantages and disadvantages

2.2 *Chemical Heat Pump System for Heat Exchange*

Chemical heat pumps operate based on fundamental principles, providing innovative ways to efficiently transfer and store thermal energy. Enhanced by Digital Twin technology, this section delves into two main types: Adsorption-type and Reaction-type chemical heat pumps. The classification and working rules of these heat pumps, simulated and analyzed through Digital Twins, are illustrated in Fig. 4 [29]. Reaction-type chemical heat pumps represent an innovative approach to thermal energy management, utilizing the principles of chemical reactions for efficient heat absorption and release [30]. Unlike adsorption-based systems, these heat pumps harness the dynamic nature of chemical transformations, offering a promising avenue for diverse applications, including synergies with geothermal energy—optimally modeled in Digital Twins.

At the heart of reaction-type chemical heat pumps are endothermic and exothermic reactions, each playing a distinctive role in the overall heat exchange process, dynamically simulated and optimized through Digital Twins [31]. Endothermic reactions, crucial to these pumps, absorb heat from their surroundings, inducing a cooling effect on the environment. This feature is invaluable for applications where localized cooling is essential, such as in industrial processes or climate control systems, with Digital Twins providing real-time control and adjustments.

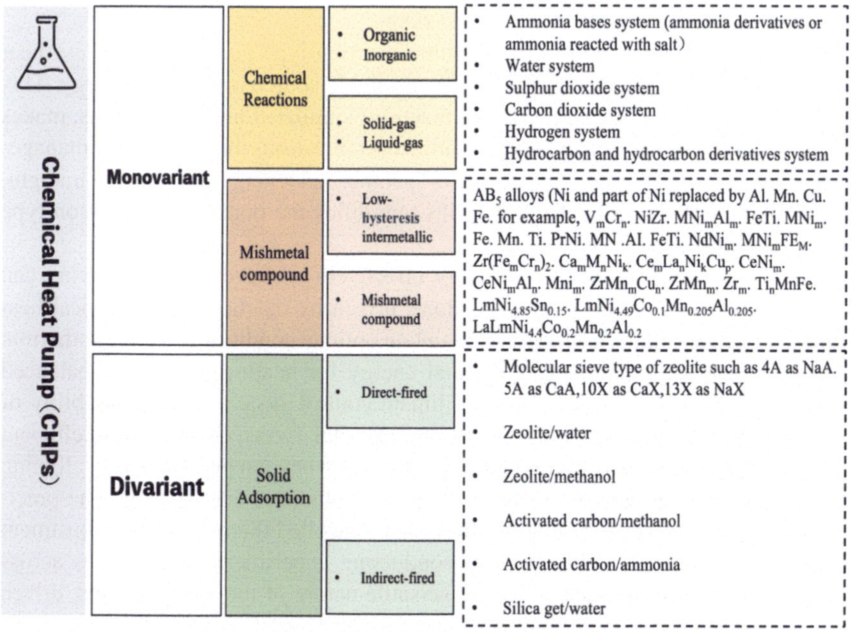

Fig. 4 Chemical heat pump classification and working rule

Adsorption-type chemical heat pumps, modeled in Digital Twins, represent a sophisticated mechanism for thermal energy transfer, rooted in the principle of adsorption [32]. This process relies on a solid material known as the adsorbent, which plays a pivotal role in selectively capturing and retaining molecules of a gas or liquid on its surface. The adsorption phase is crucial, as the adsorbent material efficiently holds the target molecules, creating a stable and controlled interaction. Digital Twin technology enables the precise control of this phase, optimizing the energy storage capacity of the system.

The process transitions into the desorption stage, where heat applied to the adsorbent prompts the release of captured molecules. This controlled release, simulated in Digital Twins, leads to the liberation of stored thermal energy, fundamental to the pump's functionality [33]. The efficiency and consistency of this phase are ensured by Digital Twin monitoring, leading to reliable thermal energy for various applications. The system then enters the regeneration stage, where the desorbed molecules, now gaseous, can be condensed or collected. Digital Twins play a crucial role in this phase, ensuring the adsorbent is ready for a new cycle of adsorption, thereby guaranteeing the continuous operation of the heat pump system [34].

The integration of geothermal energy, further enhanced by Digital Twins, boosts the capabilities of adsorption-type chemical heat pumps. By combining geothermal and adsorption technologies, a synergistic approach emerges, with Digital Twins optimizing the capture and store thermal energy from geothermal sources [35]. This integration not only maximizes the utilization of geothermal energy but also enhances the overall efficiency of the heat pump system.

Conversely, exothermic reactions within reaction-type heat pumps release heat into the surroundings, providing a readily available source of thermal energy. This dual capability, enhanced by reversible reactions simulated in Digital Twins, makes reaction-type chemical heat pumps versatile tools in various thermal energy management scenarios [36]. The integration of geothermal energy into this paradigm, modeled and optimized by Digital Twins, amplifies the potential of reaction-type chemical heat pumps [37].

In industrial applications, this synergy, effectively modeled in Digital Twins, can optimize processes. Endothermic reactions, driven by reaction-type chemical heat pumps, cool industrial equipment and maintain optimal conditions, while exothermic reactions provide supplementary thermal energy for heating, creating a balanced and energy-efficient system [38]. For climate control systems, the adaptability of reaction-type heat pumps, fine-tuned through Digital Twins, contributes to efficient heating during colder periods and effective cooling when temperatures rise, aligning with sustainable and environmentally friendly geothermal energy [39]. The precision offered by reversible reactions allows for controlled thermal output adjustment in these heat pumps, instrumental in conducting experiments and studies across various scientific disciplines [40]. The versatile nature of these heat pumps, driven by dynamic chemical reactions and enhanced by Digital Twin technology, positions them as key players in the quest for sustainable and efficient thermal energy utilization.

3 Mechanisms of Heat Fluid and Thermal Storage

In the integrated system of geothermal and chemical heat pumps, the pivotal component is the thermal fluid, which serves as the heat exchange medium. This fluid, crucial for transferring and storing energy, requires high thermal conductivity to efficiently absorb and release heat [35]. Digital Twin technology enhances this process by simulating the thermal fluid dynamics, allowing for the optimization of heat transfer between the geothermal and chemical heat pumps. This ensures that the system operates within optimal parameters for maximum efficiency. The interaction of heat fluids, pivotal for the successful integration of these technologies, is dynamically modeled and monitored using Digital Twins. Figure 5, generated from Digital Twin simulations, illustrates these interactions and the critical role of thermal fluids in enhancing system performance.

3.1 Heat Transfer Characteristics for Fluids

Optimizing energy conversion processes in geothermal and chemical heat pump systems necessitates a nuanced understanding of the intricate dynamics of heat fluid interactions, a process effectively modeled by Digital Twin technology [36]. Delving into the specific characteristics of heat fluid heat transfer unveils essential considerations that significantly impact system efficiency [37]. This includes Thermal Conductivity, Phase Change Effects, Specific Heat Capacity, Flow Dynamics, and Compatibility.

For Thermal Conductivity, various heat transfer fluids exhibit diverse thermal conductivities. Water, commonly used in geothermal applications, has a relatively high thermal conductivity [38], while specialized fluids like hydrocarbons or silicone oils are preferred in research for their enhanced properties, ensuring more efficient heat transfer. Leveraging phase change effects is paramount for enhancing energy transfer efficiency. Phase change materials (PCMs) such as paraffin wax or salt hydrates, used in chemical heat pumps, undergo transitions between solid and liquid states, absorbing or releasing significant amounts of thermal energy [39]. Digital Twins simulate these transitions to optimize heat transfer dynamically.

The specific heat capacity of heat transfer fluids determines their ability to absorb and retain thermal energy [40]. Water is frequently used in geothermal systems due to its high specific heat capacity, whereas advanced fluids like molten salts or glycols are chosen for their superior capacities, contributing to more effective energy storage [42].

In geothermal systems, ground-source heat pumps circulate water or antifreeze solutions through underground loops, taking advantage of the Earth's consistent temperature. In chemical heat pumps, the controlled flow of heat transfer fluids through reactors or heat exchangers is critical for efficient heat exchange during chemical reactions [23, 43]. Furthermore, ensuring compatibility between heat

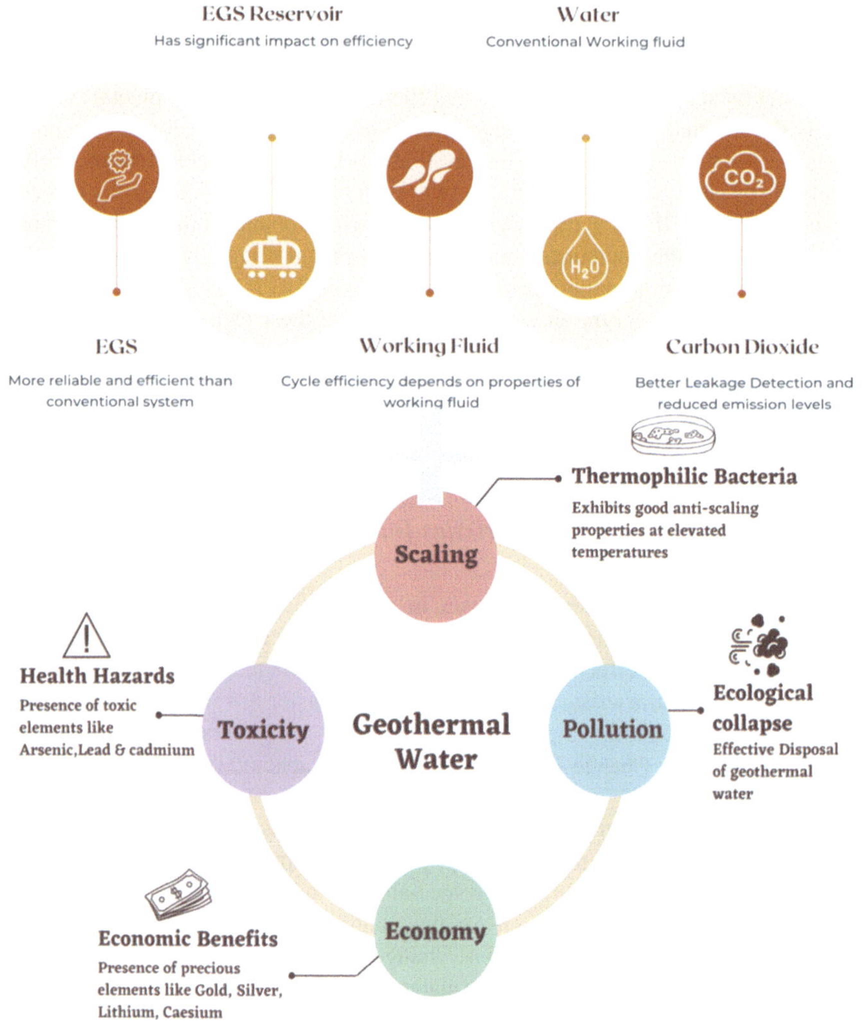

Fig. 5 Themes from geothermal energy to heat-fluid and related issues (*source* [41])

transfer fluids is imperative for system longevity. Corrosion-resistant materials, such as stainless steel, are employed to withstand the corrosive nature of some fluids. The selection of materials in contact with phase change materials considers their compatibility to prevent degradation and maintain system integrity [44]. Digital Twin technology aids in selecting and testing these materials by predicting their behavior and interaction under various operational conditions, ensuring the longevity and efficiency of the system [45].

3.2 Selection and Performance Impact of Thermal Fluids

The selection of thermal fluids plays a pivotal role in the overall performance of geothermal and chemical heat pump systems. The properties of these fluids significantly impact the efficiency, safety, and longevity of the entire energy conversion process [46]. This section delves into the criteria for selecting thermal fluids and explores how their properties, modeled through Digital Twin technology, influence system performance [47].

3.2.1 Criteria for Thermal Fluid Selection

Key criteria for thermal fluid selection in geothermal and chemical heat pump processes include thermal conductivity, chemical stability, and environmental compatibility, crucial for optimizing system performance and safety [48]. For example, propylene glycol is favored in geothermal systems for its superior heat transfer efficiency [49], while low-viscosity silicone oils optimize fluid circulation and energy consumption. High-temperature molten salts demonstrate exceptional stability under extreme conditions [50], and environmentally benign hydrofluoroether (HFE) fluids ensure system integrity [51]. Phase change materials like paraffin wax enhance energy storage and release during transitions [52], with Digital Twin technology simulating these processes to optimize fluid choices.

For sustainable solutions, selecting biodegradable and non-toxic fluids like vegetable oils aligns with environmental goals, minimizing ecological impact [53]. Digital Twin simulations assist in screening these fluids, ensuring compatibility and performance across varied operational scenarios. Figure 6 presents Thermal Fluid Screening Criteria and Typical Applications, visually enhanced by Digital Twin data.

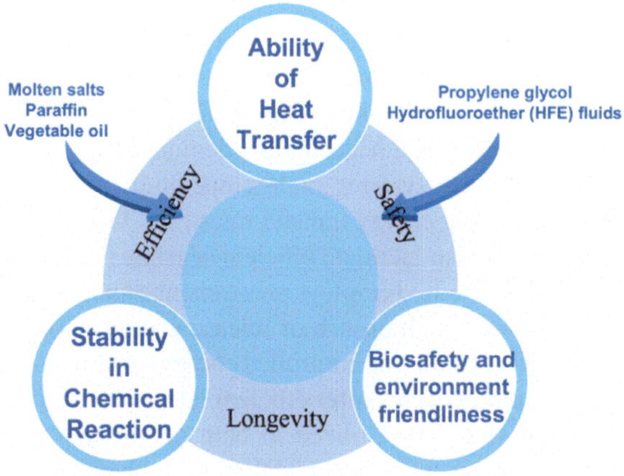

Fig. 6 Thermal fluid screening criteria and typical applications

3.3 Impact of Thermal Fluid Properties on System Performance

The properties of thermal fluids critically influence the performance and efficiency of systems involving heat transfer and temperature control [54]. These properties are essential for designing and maintaining effective geothermal and chemical processes. Thermal conductivity directly impacts energy efficiency; for example, ammonia-water mixtures used in chemical heat pump systems exhibit high thermal conductivity, enhancing heat exchange efficiency [55, 56]. Additionally, the specific heat capacity of fluids like molten salt in geothermal storage systems enables efficient energy storage and release. Viscosity influences pumping power; low-viscosity thermal oils in geothermal heating processes optimize fluid movement and reduce energy consumption.

Material compatibility is vital for system longevity, with the use of corrosion-resistant alloys and low-corrosive thermal fluids like brine solutions in geothermal power plants enhancing durability [57]. Phase change materials in air-based heat pump systems improve temperature regulation and expand operational ranges [58]. Selecting environmentally friendly thermal fluids, such as those derived from biomass, supports sustainability goals and reduces the carbon footprint of chemical heating systems [59].

Real-world applications of these thermal fluid properties help engineers design systems that prioritize efficiency, safety, and environmental sustainability, enhancing the success and longevity of advanced energy conversion technologies.

3.4 Mechanisms of Heat Fluid and Thermal Storage for Performance Improvement

Efficient heat fluid management and thermal storage mechanisms are crucial for optimizing the performance of energy conversion systems, particularly in geothermal and chemical heat pumps. This section outlines key strategies for sensible, latent, and thermochemical heat storage, demonstrating their impact on system efficiency.

In geothermal systems, high-temperature brine serves as an effective heat transfer fluid, enhancing energy transfer rates [60]. Glycol–water mixtures in chemical heat pump cycles also facilitate efficient sensible heat storage during various operational phases. This method employs phase change materials, like paraffin wax in geothermal applications, which absorb or release energy during transitions between solid and liquid states, extending controlled energy release periods [61]. Salt hydrates in chemical heat pump cycles offer efficient latent heat storage, boosting energy conversion efficiency [62, 63]. Utilized in both geothermal power plants and chemical heat pump systems, thermochemical storage involves reversible chemical reactions to store and release heat. Materials such as calcium oxide and water in geothermal applications, and metal hydrides in chemical systems, facilitate efficient thermochemical storage, enhancing system flexibility and performance [64]. Dual-phase thermal fluids that incorporate both sensible and latent heat elements optimize energy capture and release. This approach is particularly effective in chemical heat pump systems, where it enhances the overall thermal storage performance [65, 66]. Thermochemical reactions are also integrated into geothermal systems, using materials that undergo reversible chemical reactions to improve control and efficiency during energy storage and release [67].

This exploration of thermal storage mechanisms highlights their pivotal role in enhancing the efficiency and control of heat pump systems, supporting more robust and adaptable energy conversion strategies.

4 Heat Fluid Screening for Thermal Storage Optimization

4.1 Selection of Thermal Fluids for Increase Heat Storage

The selection of appropriate thermal fluids is critical for optimizing thermal storage within energy systems, influencing the efficiency and effectiveness of storage processes. Thermal fluids with high thermal conductivity facilitate rapid heat transfer, enhancing the absorption and release of thermal energy, thus improving thermal storage capabilities [68]. Phase Change Materials (PCMs) undergo phase transitions like melting and solidification, significantly boosting thermal storage efficiency by absorbing and releasing latent heat [69]. This characteristic makes them suitable for use as thermal fluids, as shown in classifications and enthalpy ranges in Fig. 7.

The specific heat capacity of a fluid determines its energy storage potential. Fluids with a higher specific heat capacity can store more energy, thereby increasing the

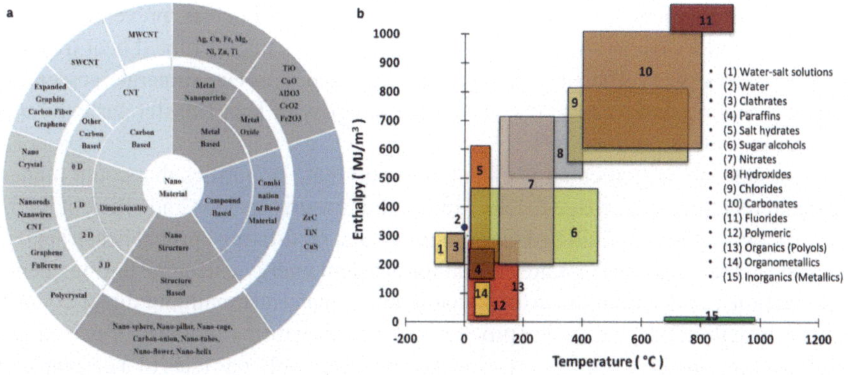

Fig. 7 **a** Classifications of PCM nanomaterials in heat transfer and **b** Enthalpy and temperature ranges (*data source* [70, 71])

system's thermal storage capacity [72]. Thermal fluids also need to be stable and compatible with system materials across varying temperatures to ensure long-term reliability [73]. Low viscosity is crucial for promoting smooth fluid circulation and efficient heat transfer within the system [74].

Environmentally friendly thermal fluids, such as biodegradable options, align with sustainability goals, reducing the environmental impact of energy systems [75]. The operational temperature range of the fluid should match the system's requirements to ensure effective performance under diverse conditions [76]. Cost-effectiveness is also a consideration; selecting affordable fluids that meet performance standards supports the economic feasibility of the system [77].

Enhanced thermal storage via suitable fluids helps match energy production with demand, stabilizing system operation and minimizing energy wastage by acting as a buffer for excess thermal energy [78, 79]. This load matching capability and the ability to maintain stable temperatures during chemical reactions optimize the efficiency of both geothermal and chemical heat pump systems [80].

By ensuring better integration with intermittent renewable energy sources and providing flexibility in energy supply, thermal storage supports grid stability and enhances the system's responsiveness to changes in energy demand [81–85]. Careful consideration of these criteria when selecting thermal fluids maximizes thermal storage capacity, contributing to the overall efficiency and effectiveness of energy systems.

4.2 *Heat Fluid Selection for Improving Storage Efficiency*

Heat fluid screening is crucial for optimizing heat transfer processes and enhancing thermal storage efficiency in various systems. This section details how strategic screening contributes to system performance by maintaining the purity and efficacy of heat fluids.

Screening removes contaminants such as sediment and debris, preventing clogging in ground loops and heat exchangers. This ensures uninterrupted heat transfer and maintains system efficiency [86]. By capturing and removing minerals, screening mitigates scaling and fouling, preserving the integrity of heat transfer surfaces and preventing scale formation [87].

In these systems, screening is essential to eliminate impurities that could hinder chemical reactions or damage reactors [88, 89]. It ensures a clean and efficient interaction between the heat fluid and reactive materials, facilitating optimal heat transfer. Clean heat exchange surfaces are crucial for efficient energy exchange during chemical reactions, and screening ensures that reactive materials maintain direct contact with the heat fluid [87]. Screening processes also control the particle size of suspended solids, optimizing heat transfer efficiency by ensuring only particles of a certain size pass through [90]. This is particularly critical in chemical heat pump systems to prevent blockages and undesired reactions, ensuring the heat fluid remains free from particles that could interfere with processes [91]. Regular screening of heat fluids enhances the longevity and reliability of both geothermal and chemical heat pump systems. By preventing the accumulation of contaminants and maintaining clean heat exchange surfaces, screening supports sustained system performance and efficiency over time.

By maintaining the cleanliness and appropriate composition of heat fluids, screening plays a pivotal role in enhancing the overall efficiency and performance of energy systems, contributing to superior thermal storage capabilities.

4.3 Optimizing Methods for Enhancing Heat Storage Through Heat Fluid Selection

Efficient thermal energy storage in geothermal and chemical heat pump systems necessitates a comprehensive approach to optimize heat fluid screening. Several strategies are strategically integrated to enhance the screening process, ensuring optimal performance and prolonged system longevity [92]. Implementing advanced filtration systems that consider various parameters, including particle size, chemical composition, and viscosity, forms a cornerstone for comprehensive heat fluid screening. Smart filtration algorithms dynamically adjust parameters in real-time, maintaining high-quality heat fluids and preventing potential system issues. Machine learning applications play a crucial role in predictive maintenance, allowing for the anticipation of issues in heat fluids. This proactive approach enables timely interventions, preserving system efficiency [93]. Optimization algorithms, driven by machine learning, analyze system parameters, suggesting adjustments to enhance heat transfer efficiency and guide the selection of optimal heat fluids. Advanced material characterization, coupled with nano-material innovations, contributes to the continuous assessment of thermal material conditions [94]. This facilitates adaptive responses, ensuring stability and longevity. The exploration of nanotechnology enhances heat transfer properties, improving conductivity and resistance to degradation. Dynamic system modeling, particularly through Computational Fluid Dynamics (CFD) simulations, provides insights into heat fluid behavior and thermal material interactions [95]. By optimizing fluid flow patterns and identifying potential bottlenecks, CFD simulations guide the selection of materials with superior efficiency and stability [96]. Classifications of methods for enhancing heat storage through heat fluid selection has been shown in Fig. 8.

The methods for optimizing heat storage through the manipulation of thermal fluids can typically be categorized into experimental methods and physical modeling approaches: On the one hand, experimental methods involve measuring the thermal properties of fluids to assess their efficiency in storing and releasing heat. This often entails utilizing devices like Differential Scanning Calorimeters (DSC) to measure the specific heat capacity of fluids [97]. Additionally, assessing the thermal conductivity of fluids is crucial for optimizing heat storage systems, which can be achieved through methods such as thermal conductivity meters or the hot plate method. Evaluating the

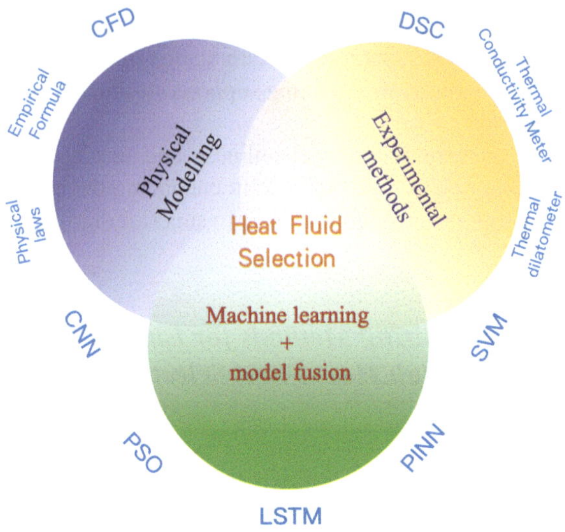

Fig. 8 Classifications of methods for enhancing heat storage through heat fluid selection

stability and longevity of fluid circulation in storing and releasing heat under real-world conditions is another essential experimental method. This can be accomplished by simulating the cyclic storage and release of heat and assessing the fluid's stability and lifespan. Understanding the dynamic thermal behavior of fluids under various temperatures and pressures can also be achieved through techniques like dynamic thermal analysis using instruments such as thermal expansion analyzers or dynamic mechanical analyzers.

On the other hand, physical modeling involves creating mathematical representations of fluid dynamics and heat transfer phenomena. Computational Fluid Dynamics (CFD) is commonly used to simulate fluid flow and heat transfer behaviors within thermal storage systems, aiding in optimization efforts [98]. Developing physical models that account for the interactions between heat and fluid transport can help evaluate and select fluids for optimal performance in thermal storage processes [99]. Similarly, establishing mathematical models based on fluid dynamics and heat conduction principles allows for predicting the stability and efficiency of fluid circulation in storing and releasing heat [100]. Furthermore, for systems involving multiple fluid phases or phase transitions, appropriate multiphase flow models can be developed to capture the complex behavior of fluids, thereby optimizing the performance of thermal storage systems [101].

The integration of these experimental methods and physical modeling techniques provides a comprehensive theoretical foundation and experimental validation for optimizing the application of fluids in thermal storage systems [102]. Moreover, further optimization approaches, such as leveraging machine learning, neural networks, or physics-informed neural networks, can be applied in conjunction with the aforementioned methods to enhance and refine heat storage optimization efforts. The integration of geothermal and chemical heat pump technologies in a hybrid system forms another critical aspect [103]. This approach leverages the strengths of both technologies, optimizing thermal energy storage, and enhancing overall system efficiency. Designing the system with modularity and flexibility allows seamless integration of new technologies or materials, ensuring adaptability to advancements in heat fluid screening and thermal material innovations. Collectively, these strategies create a holistic and interconnected approach to optimizing heat fluid screening. This comprehensive framework ensures superior heat storage efficiency, meeting the demands of modern energy systems while contributing to sustainable and effective energy conversion.

5 Results and Discussion

To conduct a meta-analysis on chemical heat pumps in geothermal contexts, we reviewed literature from 2010 to 2023, noting a significant increase in publications from 5,910 to 21,071, indicating a growing research interest (Fig. 9a). A rigorous screening process was applied to refine this vast body of work, focusing on articles indexed in the Science Citation Index (SCI) or SCI Expanded, excluding review articles and those lacking specific experimental data on the application in geothermal energy. This ensured that only the most relevant and informative studies were included in the analysis.

The rigorous screening criteria ensured that only relevant and informative literature was included in the meta-analysis, focusing on chemical heat pumps in geothermal contexts. Significant interest in this intersection is evidenced by the extensive number of publications depicted in Fig. 9b, c. Notably, solid–gas and low-hysteresis intermetallic types of chemical heat pumps drew considerable attention, with 58 and 45 papers, respectively, highlighting a wide array of research focuses and the potential for innovation in the field.

Among the various types of geothermal technologies, flash steam geothermal power generation emerged as the most researched, with 286 articles, while liquid–gas chemical heat pump studies dominated chemical pump research with 86 articles. However, the research landscape shows notable gaps; for example, only 17 articles addressed indirect-fired chemical heat pumps, and dry steam geothermal power generation was also underrepresented with just 48 articles. These findings point to the need for more comprehensive research efforts to deepen understanding and foster advancement across all facets of geothermal and chemical heat pump technologies.

Based on COP data analyzed through Digital Twin simulations, distinct performance characteristics of various thermal working fluids are evident, as depicted in Fig. 10. Ammonia maintains a stable average COP of 2.77, ideal for medium to high-temperature applications. Water closely follows with an average COP of 2.48, suitable for similar temperature ranges. Hydrocarbons achieve a slightly higher average COP of 2.68, indicating their stability and suitability for similar uses. Silicone oil, with an average COP of 2.64, also performs effectively in high-temperature applications.

Fluorocarbons, exhibiting the highest average COP at 3.19, are optimal for high-temperature and specialized environments. However, using Digital Twin technology,

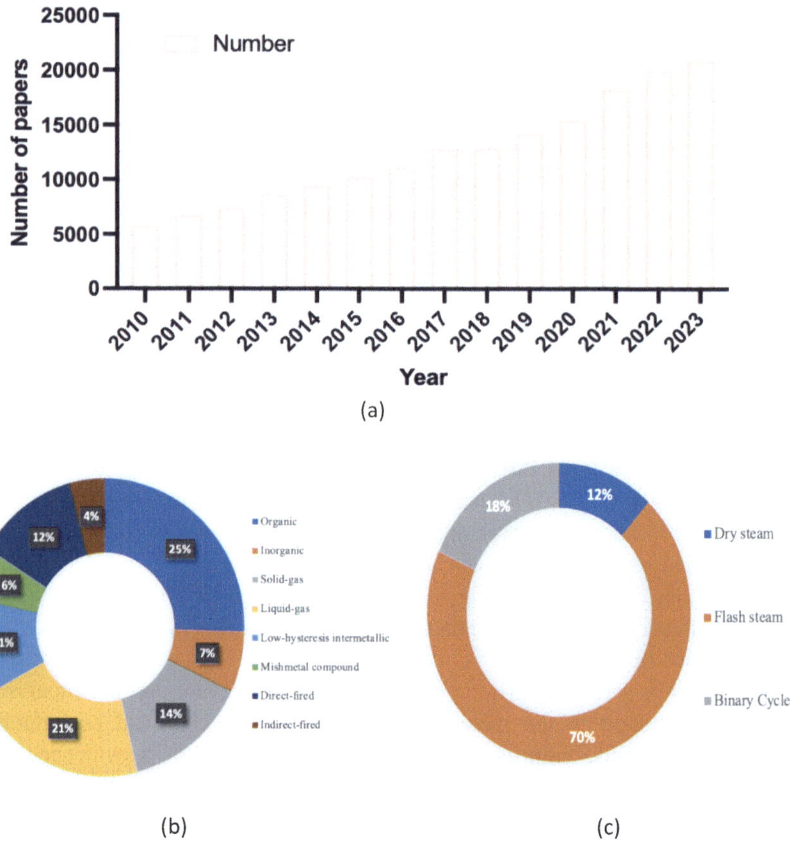

Fig. 9 Selected papers analysis after selection: **a** Papers of chemical heat pump in recent years; **b** Paper division of different chemical heat pumps in coupled system; **c** Paper division based on geothermal plants of coupled system

the environmental concerns associated with fluorocarbons—such as high Global Warming Potential (GWP) and Ozone Depletion Potential (ODP)—can be dynamically modeled to assess their overall sustainability impact. Nanofluids, with an average COP of 2.95, demonstrate versatility across different temperature ranges, further supported by Digital Twin simulations to optimize their application based on real-time environmental and performance data.

There are variations in the Levelized Cost of Heat (LCOH) among different thermal working fluids, as shown in Fig. 11. Ammonia exhibits a relatively stable LCOH ranging from 0.05 to 0.1 USD per MJ, making it suitable for medium to high-temperature applications. Water shows a narrower range, with LCOH fluctuating between 0.03 to 0.08 USD per MJ, representing a cost-effective option. Hydrocarbons demonstrate a broader range of LCOH, from 0.06 to 0.15 USD per MJ, indicating variability but still remaining a common and economical choice. Silicone

Fig. 10 COP for different heat fluids in chemical-geothermal heat pump

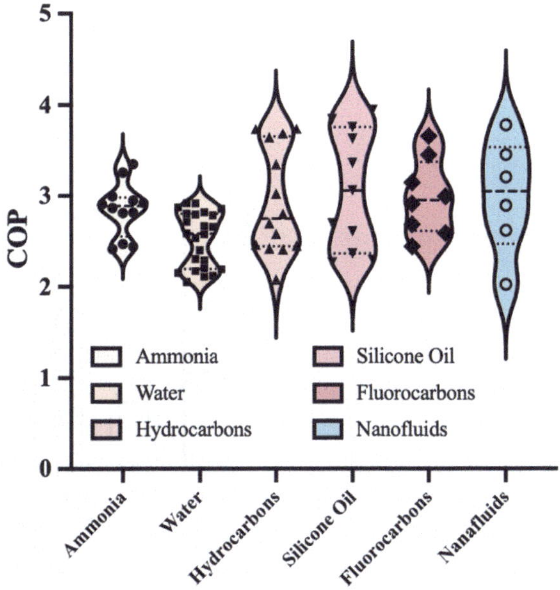

oil shows fluctuation between 0.08 to 0.15 USD per MJ, possibly influenced by factors such as investment costs and operational efficiency. Fluorocarbons present the highest LCOH, ranging from 0.1 to 0.4 USD per MJ, primarily due to their high investment costs and environmental concerns. Nanofluids display considerable variability in data, with significant differences between upper and lower limits, requiring further research to determine their actual LCOH range. In summary, water emerges as a cost-effective option, while ammonia and hydrocarbons remain common and economically feasible choices. However, fluorocarbons may face challenges due to their high costs and environmental issues, and nanofluids require more investigation to ascertain their economic viability. When selecting thermal working fluids, factors such as cost, performance, and environmental impact should be carefully considered.

The Meta-analysis reveals a growing interest in chemical heat pump technology within the geothermal context, with a significant number of 409 publications indicating a diverse range of research focuses and interests. Among thermal working fluids, while fluorocarbons demonstrate the highest COP values, which even over 4.5, their environmental concerns necessitate careful consideration. Water emerges as a cost-effective option, while ammonia and hydrocarbons remain common choices for medium to high-temperature applications. The Levelized Cost of Heat (LCOH) varies among different thermal working fluids, with water presenting as a cost-effective option, with only 0.12 USD per MJ, while fluorocarbons exhibit higher costs primarily due to environmental concerns, underscoring the importance of considering economic viability alongside performance and environmental impact when selecting thermal working fluids.

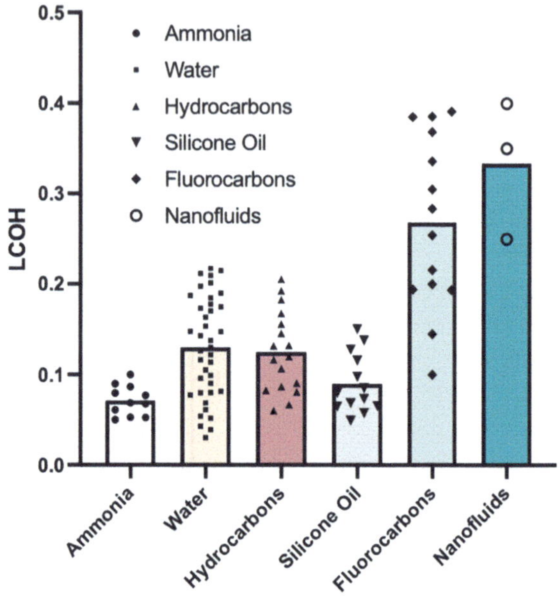

Fig. 11 LCOH for different heat fluids in chemical-geothermal heat pump

6 Conclusion

Chemical heat pump systems, integrated with geothermal energy sources, offer a sustainable solution for efficient thermal energy conversion and utilization. These systems benefit significantly from the application of Digital Twin technology, which simulates and optimizes the operation and performance of both geothermal and chemical components.

Key contributions of this paper include:

(1) A thorough introduction to the principles, mechanisms, and applications of geothermal and chemical heat pump systems, enhanced by Digital Twin simulations that visualize and predict system behaviors under various operational scenarios.
(2) A detailed discussion on the critical role of thermal fluids, where Digital Twin technology aids in selecting optimal fluids by analyzing their heat transfer characteristics and storage capacities in real time.
(3) A meta-analysis that identifies a growing interest in integrating chemical heat pumps with geothermal energy, supported by Digital Twin data showing high COP values for fluorocarbons and the cost-effectiveness of water. The analysis also points to emerging areas such as organic chemical heat pumps and nanofluids, where further detailed research is needed.
(4) Identification of challenges and potential advancements in geothermal-chemical heat pump systems, with Digital Twin models providing insights into addressing these challenges and improving system performance.

By systematically leveraging Digital Twin technology, this paper offers a comprehensive understanding of geothermal-chemical heat pump systems, providing valuable references for future research and practical applications. This approach not only enhances the overview and discussion of optimization methods but also emphasizes the transformative impact of Digital Twin technology in advancing thermal management solutions.

References

1. Hou G, Xu L, Liu Z, Chen D, Ru H, Taherian H (2023) Chapter 6—Solar-assisted geothermal heat pump systems: current practice and future development. In: Jeguirim M, Dutournié P (eds) Renewable energy production and distribution, vol 2. Academic Press, pp 217–246
2. Hou G, Liu Z, Zhao M, Taherian H, Jiang W, Chen D (2021) Chapter 6—Underground energy: utilization of geothermal shallow heat pumps. In: Jeguirim M (ed) Recent advance in renewable energy technologies. Academic Press, pp 211–247
3. Liu Z, Chen Y, Yang X, Yan J (2023) Power to heat: opportunity of flexibility services provided by building energy systems. Adv Appl Energy 11:100149. https://doi.org/10.1016/j.adapen.2023.100149
4. Liu Z, Hou G, Song Y, Taherian H, Qi S The impact of soil hydrothermal properties on geothermal power generation (GPG): modeling and analysis. Energies 15(2). https://doi.org/10.3390/en15020448
5. Soltani M, Moradi Kashkooli F, Alian Fini M, Gharapetian D, Nathwani J, Dusseault MB (2022) A review of nanotechnology fluid applications in geothermal energy systems. J Renew Sustain Energy Rev 167:112729. https://doi.org/10.1016/j.rser.2022.112729
6. Szabó GL Thermo-chemical instability and energy analysis of absorption heat pumps. Energies 13(8). https://doi.org/10.3390/en13081966
7. Razaghi M, Rahdar MH, Aghamohamadi N, Mohammadi M, Ahmadi P (2024) Biofuel, solar-biofuel or geothermal energy; which resource can better contribute to an integrated energy system for residential energy demands. J Process Saf Environ Prot 181:103–130. https://doi.org/10.1016/j.psep.2023.10.063
8. Marina A, Spoelstra S, Zondag HA, Wemmers AK (2021) An estimation of the European industrial heat pump market potential. J Renew Sustain Energy Rev, 139:110545. https://doi.org/10.1016/j.rser.2020.110545
9. Jiang J, Hu B, Wang RZ, Deng N, Cao F, Wang C-C (2022) A review and perspective on industry high-temperature heat pumps. J Renew Sustain Energy Rev 161:112106. https://doi.org/10.1016/j.rser.2022.112106
10. Kiss AA, Smith R (2020) Rethinking energy use in distillation processes for a more sustainable chemical industry. Energy 203:117788. https://doi.org/10.1016/j.energy.2020.117788
11. Topuz A, Engin T, Erdoğan B, Mert S, Yeter A (2020) Experimental investigation of pressure drop and cooling performance of an automobile radiator using Al2O3-water + ethylene glycol nanofluid. Heat Mass Transf 56(10):2923–2937. https://doi.org/10.1007/s00231-020-02916-8
12. Kang W, Mushi SJ, Yang H, Wang P, Hou X (2020) Development of smart viscoelastic surfactants and its applications in fracturing fluid: a review. J Pet Sci Eng 190:107107. https://doi.org/10.1016/j.petrol.2020.107107
13. Chen H, Liu H, Zhang S, Feng Y (2020) Smart thermoviscosifying polymer for improving drag reduction in slick-water hydrofracking. Fuel 278:118408. https://doi.org/10.1016/j.fuel.2020.118408
14. Yadav K et al (2023) Geothermal-solar integrated Multistage flash distillation cogeneration system: a cleaner and sustainable solution. Desalination 566:116897. https://doi.org/10.1016/j.desal.2023.116897

15. Gao MW et al (2023) A novel triple responsive smart fluid for tight oil fracturing-oil expulsion integration. J Pet Sci 20(2):982–992. https://doi.org/10.1016/j.petsci.2023.01.008
16. Muther T, Syed FI, Lancaster AT, Salsabila FD, Dahaghi AK, Negahban S (2022) Geothermal 4.0: AI-enabled geothermal reservoir development—current status, potentials, limitations, and ways forward. J Geotherm 100:102348. https://doi.org/10.1016/j.geothermics.2022.102348
17. Fadhel MI, Sopian K, Daud WRW, Alghoul MA (2011) Review on advanced of solar assisted chemical heat pump dryer for agriculture produce. J Renew Sustain Energy Rev 15(2):1152–1168. https://doi.org/10.1016/j.rser.2010.10.007
18. Chan CW, Ling-Chin J, Roskilly AP (2013) A review of chemical heat pumps, thermodynamic cycles and thermal energy storage technologies for low grade heat utilization. J Appl Therm Eng 50(1):1257–1273. https://doi.org/10.1016/j.applthermaleng.2012.06.041
19. Karaca F, Kıncay O, Bolat E (2002) Economic analysis and comparison of chemical heat pump systems. J Appl Therm Eng 22(16):1789–1799. https://doi.org/10.1016/S1359-4311(02)00078-9
20. Deng J, Wei Q, He S, Liang M, Zhang H Simulation analysis on the heat performance of deep borehole heat exchangers in medium-depth geothermal heat pump systems. Energies 13(3). https://doi.org/10.3390/en13030754
21. Kato Y, Yamashita N, Kobayashi K, Yoshizawa Y (1996) Kinetic study of the hydration of magnesium oxide for a chemical heat pump. J Appl Therm Eng 16(11):853–862. https://doi.org/10.1016/1359-4311(96)00009-9
22. Wang Y, Liu Y, Dou J, Li M, Zeng M (2020) Geothermal energy in China: status, challenges, and policy recommendations. Utilities Policy 64:101020. https://doi.org/10.1016/j.jup.2020.101020
23. Liu Z, Yang X, Ali HM, Liu R, Yan J (2023) Multi-objective optimizations and multi-criteria assessments for a nanofluid-aided geothermal PV hybrid system. Energy Rep 9:96–113. https://doi.org/10.1016/j.egyr.2022.11.170
24. Wang L, Huang X, Babaei M, Liu Z, Yang X, Yan J (2023) Full-scale utilization of geothermal energy: a high-efficiency CO_2 hybrid cogeneration system with low-temperature waste heat. J Clean Prod 403:136866. https://doi.org/10.1016/j.jclepro.2023.136866
25. Moeck IS (2014) Catalog of geothermal play types based on geologic controls. J Renew Sustain Energy Rev 37:867–882. https://doi.org/10.1016/j.rser.2014.05.032
26. Hähnlein S, Bayer P, Ferguson G, Blum P (2013) Sustainability and policy for the thermal use of shallow geothermal energy. Energy Policy 59:914–925. https://doi.org/10.1016/j.enpol.2013.04.040
27. Zhang L, Wang Y, Ding B, Gu J, Ukrainczyk N, Cai J (2023) Development of geopolymer-based composites for geothermal energy applications. J Clean Prod 419:138202. https://doi.org/10.1016/j.jclepro.2023.138202
28. Mahmoud M, Ramadan M, Naher S, Pullen K, Ali Abdelkareem M, Olabi AG (2021) A review of geothermal energy-driven hydrogen production systems. J Therm Sci Eng Prog 22:100854. https://doi.org/10.1016/j.tsep.2021.100854
29. de Boer R, Haije WG, Veldhuis JBJ (2002) Determination of structural, thermodynamic and phase properties in the Na2S–H2O system for application in a chemical heat pump J Thermochim Acta 395(1):3–19. https://doi.org/10.1016/S0040-6031(02)00158-2
30. Obermeier J, Müller K, Arlt W (2015) Thermodynamic analysis of chemical heat pumps. Energy 88:489–496. https://doi.org/10.1016/j.energy.2015.05.076
31. Fujimoto S, Bilgen E, Ogura H (2002) CaO/Ca(OH)2 chemical heat pump system. J Energy Convers Manage 43(7):947–960. https://doi.org/10.1016/S0196-8904(01)00081-4
32. He D, Tang X, Rehan MA, Lu Y, Li G (2024) A novel high-efficient continuous power generation device employing thermally regenerative electrochemical cycle. J Appl Therm Eng 236:121879. https://doi.org/10.1016/j.applthermaleng.2023.121879
33. Dai S, Yin Y, Li W, Zhang F (2020) Thermodynamic analysis of a novel chemical heat pump cycle based on the physical-chemical thermal effects of reversible reaction. J Energy Convers Manage 225:113419. https://doi.org/10.1016/j.enconman.2020.113419

34. Gondal IA (2021) Prospects of shallow geothermal systems in HVAC for NZEB. J Energy Built Environ 2(4):425–435. https://doi.org/10.1016/j.enbenv.2020.09.007
35. Palomo-Torrejón E, Colmenar-Santos A, Rosales-Asensio E, Mur-Pérez F (2021) Economic and environmental benefits of geothermal energy in industrial processes. Renewable Energy 174:134–146. https://doi.org/10.1016/j.renene.2021.04.074
36. Ramesh KN, Sharma TK, Rao GAP (2021) Latest advancements in heat transfer enhancement in the micro-channel heat sinks: a review. Archiv Comput Methods Eng 28(4):3135–3165. https://doi.org/10.1007/s11831-020-09495-1
37. Ali ARI, Salam B (2020) A review on nanofluid: preparation, stability, thermophysical properties, heat transfer characteristics and application. J SN Appl Sci 2(10):1636. https://doi.org/10.1007/s42452-020-03427-1
38. Krishna Y, Faizal M, Saidur R, Ng KC, Aslfattahi N (2020) State-of-the-art heat transfer fluids for parabolic trough collector. Int J Heat Mass Transf 152:119541. https://doi.org/10.1016/j.ijheatmasstransfer.2020.119541
39. Sarada K, Gowda RJP, Sarris IE, Kumar RN, Prasannakumara BC Effect of magnetohydrodynamics on heat transfer behaviour of a non-newtonian fluid flow over a stretching heat under local thermal non-equilibrium condition. Fluids 6(8). https://doi.org/10.3390/fluids6080264
40. Pandya NS, Shah H, Molana M, Tiwari AK (2020) Heat transfer enhancement with nanofluids in plate heat exchangers: A comprehensive review. Eur J Mech B/Fluids 81:173–190. https://doi.org/10.1016/j.euromechflu.2020.02.004
41. Rohit RV et al (2023) Tracing the evolution and charting the future of geothermal energy research and development. Renew Sustain Energy Rev 184:113531. https://doi.org/10.1016/j.rser.2023.113531
42. Soltani M et al (2021) Environmental, economic, and social impacts of geothermal energy systems. Renew Sustain Energy Rev 140:110750. https://doi.org/10.1016/j.rser.2021.110750
43. Radhika M, Punith Gowda RJ, Naveenkumar R, Siddabasappa, Prasannakumara BC (2021) Heat transfer in dusty fluid with suspended hybrid nanoparticles over a melting surface. Heat Transf 50(3):2150–2167. https://doi.org/10.1002/htj.21972
44. Zhang X (2021) Numerical study of geothermal district heating from a ground heat exchanger coupled with a heat pump system. Appl Therm Eng 185:116335. https://doi.org/10.1016/j.applthermaleng.2020.116335
45. Wang G, Chen T, Tian M, Ding G (2020) Fluid and heat transfer characteristics of microchannel heat sink with truncated rib on sidewall. Int J Heat Mass Transf 148:119142. https://doi.org/10.1016/j.ijheatmasstransfer.2019.119142
46. Fadhel MI, Sopian K, Daud WRW (2010) Performance analysis of solar-assisted chemical heat-pump dryer. Solar Energy 84(11):1920–1928. https://doi.org/10.1016/j.solener.2010.07.001
47. Xia Q et al (2023) Efficient distributed concentrating solar power system with ammonia-based chemical heat pump. Energy Conver Manage 276:116575. https://doi.org/10.1016/j.enconman.2022.116575
48. Mbaye M, Aidoun Z, Valkov V, Legault A (1998) Analysis of chemical heat pumps (CHPs): basic concepts and numerical model description. Appl Therm Eng 18(3):131–146. https://doi.org/10.1016/S1359-4311(97)00027-6
49. Han X, Chen X, Sun Y, Qu J (2020) Performance improvement of a PV/T system utilizing Ag/CoSO4-propylene glycol nanofluid optical filter. Energy 192:116611. https://doi.org/10.1016/j.energy.2019.116611
50. Roper R et al (2022) Molten salt for advanced energy applications: a review. Ann Nucl Energy 169:108924. https://doi.org/10.1016/j.anucene.2021.108924
51. Jovell D, Gonzalez-Olmos R, Llovell F (2022) A computational drop-in assessment of hydrofluoroethers in organic Rankine cycles. Energy 254:124319. https://doi.org/10.1016/j.energy.2022.124319
52. Sun H, Khalifa NT, Saad HA, Othman HA, Hussin AM, Helmi MM (2022) Investigation of behavior of heat storage system including paraffin with involving MWCNT nanoparticles. J Energy Storage 54:105250. https://doi.org/10.1016/j.est.2022.105250

53. Barai DP, Bhanvase BA, Sonawane SH (2020) A review on graphene derivatives-based nanofluids: investigation on properties and heat transfer characteristics. Ind Eng Chem Res 59(22):10231–10277. https://doi.org/10.1021/acs.iecr.0c00865
54. Torabi M, Karimi N, Torabi M, Peterson GP, Simonson CJ (2020) Generation of entropy in micro thermofluidic and thermochemical energy systems—a critical review. Int J Heat Mass Transf 163:120471. https://doi.org/10.1016/j.ijheatmasstransfer.2020.120471
55. Sharma P et al Recent advances in machine learning research for nanofluid-based heat transfer in renewable energy system. Energy Fuels 36(13):6626–6658. https://doi.org/10.1021/acs.energyfuels.2c01006
56. Hemmat Esfe M, Bahiraei M, Hajbarati H, Valadkhani M (2020) A comprehensive review on convective heat transfer of nanofluids in porous media: energy-related and thermohydraulic characteristics. Appl Therm Eng 178:115487. https://doi.org/10.1016/j.applthermaleng.2020.115487
57. Luo J, Zou D, Wang Y, Wang S, Huang L (2022) Battery thermal management systems (BTMs) based on phase change material (PCM): a comprehensive review. Chem Eng J 430:132741. https://doi.org/10.1016/j.cej.2021.132741
58. Ran F, Chen Y, Cong R, Fang G (2020) Flow and heat transfer characteristics of microencapsulated phase change slurry in thermal energy systems: a review. Renew Sustain Energy Rev 134:110101. https://doi.org/10.1016/j.rser.2020.110101
59. Wu W et al (2020) A novel composite PCM for seasonal thermal energy storage of solar water heating system. Renew Energy 161:457–469. https://doi.org/10.1016/j.renene.2020.06.147
60. Feng PH, Zhao BC, Wang RZ (2020) Thermophysical heat storage for cooling, heating, and power generation: a review. Appl Therm Eng 166:114728. https://doi.org/10.1016/j.applthermaleng.2019.114728
61. Sheikholeslami M, Keshteli AN, Babazadeh H (2020) Nanoparticles favorable effects on performance of thermal storage units. J Mol Liq 300:112329. https://doi.org/10.1016/j.molliq.2019.112329
62. Eppinger B, Zigan L, Karl J, Will S (2020) Pumped thermal energy storage with heat pump-ORC-systems: comparison of latent and sensible thermal storages for various fluids. Appl Energy 280:115940. https://doi.org/10.1016/j.apenergy.2020.115940
63. Fernández AG, Galleguillos H, Pérez FJ (2014) Corrosion ability of a novel heat transfer fluid for energy storage in CSP plants. Oxid Metals 82(5):331–345. https://doi.org/10.1007/s11085-014-9494-3
64. Almahdi M, Dincer I, Rosen MA (2016) A new solar based multigeneration system with hot and cold thermal storages and hydrogen production. Renew Energy 91:302–314. https://doi.org/10.1016/j.renene.2016.01.069
65. Benato A, Stoppato A (2018) Heat transfer fluid and material selection for an innovative pumped thermal electricity storage system. Energy 147:155–168. https://doi.org/10.1016/j.energy.2018.01.045
66. Agyenim F, Hewitt N, Eames P, Smyth M (2010) A review of materials, heat transfer and phase change problem formulation for latent heat thermal energy storage systems (LHTESS). Renew Sustain Energy Rev 14(2):615–628. https://doi.org/10.1016/j.rser.2009.10.015
67. Liang D, Bai W, Chen W, Chyu MK (2020) Investigating the effect of element shape of the face-centered cubic lattice structure on the flow and endwall heat transfer characteristics in a rectangular channel. Int J Heat Mass Transf 153:119579. https://doi.org/10.1016/j.ijheatmasstransfer.2020.119579
68. Ahmed N, Elfeky KE, Lu L, Wang QW (2020) Thermal performance analysis of thermocline combined sensible-latent heat storage system using cascaded-layered PCM designs for medium temperature applications. Renew Energy 152:684–697. https://doi.org/10.1016/j.renene.2020.01.073
69. Guerraiche D, Bougriou C, Guerraiche K, Valenzuela L, Driss Z (2020) Experimental and numerical study of a solar collector using phase change material as heat storage. J Energy Storage 27:101133. https://doi.org/10.1016/j.est.2019.101133

70. Choure BK, Alam T, Kumar R (2023) A review on heat transfer enhancement techniques for PCM based thermal energy storage system. J Energy Storage 72:108161. https://doi.org/10.1016/j.est.2023.108161
71. Fallahi A, Guldentops G, Tao M, Granados-Focil S, Van Dessel S (2017) Review on solid-solid phase change materials for thermal energy storage: molecular structure and thermal properties. Appl Therm Eng 127:1427–1441. https://doi.org/10.1016/j.applthermaleng.2017.08.161
72. Amy C, Kelsall CC, LaPotin A, Pishahang M, Henry A (2021) Chapter 3—ultrahigh temperature sensible heat storage and heat transfer fluids. In: Datas A (ed) Ultra-high temperature thermal energy storage, transfer and conversion. Woodhead Publishing, pp 57–84
73. Frate GF, Ferrari L, Desideri U (2020) Multi-criteria investigation of a pumped thermal electricity storage (PTES) system with thermal integration and sensible heat storage. Energy Convers Manage 208:112530. https://doi.org/10.1016/j.enconman.2020.112530
74. Giambastiani BMS, Tinti F, Mendrinos D, Mastrocicco M (2014) Energy performance strategies for the large scale introduction of geothermal energy in residential and industrial buildings: the GEO.POWER project. Energy Policy 65:315–322. https://doi.org/10.1016/j.enpol.2013.10.008
75. Nabavi M, Nazarpour V, Alibak AH, Bagherzadeh A, Alizadeh SM (2021) Smart tracking of the influence of alumina nanoparticles on the thermal coefficient of nanosuspensions: application of LS-SVM methodology. Appl Nanosci 11(7):2113–2128. https://doi.org/10.1007/s13204-021-01949-7
76. Pagkalos C, Dogkas G, Koukou MK, Konstantaras J, Lymperis K, Vrachopoulos MG (2020) Evaluation of water and paraffin PCM as storage media for use in thermal energy storage applications: a numerical approach. Int J Thermofluids 1–2:100006. https://doi.org/10.1016/j.ijft.2019.100006
77. Liu Z, Yang X, Jia W, Li H, Yang X (2020) Justification of CO_2 as the working fluid for a compressed gas energy storage system: a thermodynamic and economic study. J Energy Storage 27:101132. https://doi.org/10.1016/j.est.2019.101132
78. Javadi FS, Metselaar HSC, Ganesan P (2020) Performance improvement of solar thermal systems integrated with phase change materials (PCM), a review. Solar Energy 206:330–352. https://doi.org/10.1016/j.solener.2020.05.106
79. Chauhan VK, Shukla SK, Rathore PKS (2022) A systematic review for performance augmentation of solar still with heat storage materials: a state of art. J Energy Storage 47:103578. https://doi.org/10.1016/j.est.2021.103578
80. Aftab W, Usman A, Shi J, Yuan K, Qin M, Zou R (2021) Phase change material-integrated latent heat storage systems for sustainable energy solutions. Energy Environ Sci. 14(8):4268–4291. https://doi.org/10.1039/D1EE00527H
81. Sheikholeslami M (2022) Analyzing melting process of paraffin through the heat storage with honeycomb configuration utilizing nanoparticles. J Energy Storage 52:104954. https://doi.org/10.1016/j.est.2022.104954
82. Vengadesan E, Bharathwaj D, Kumar BS, Senthil R (2022) Experimental study on heat storage integrated flat plate solar collector for combined water and air heating in buildings. Appl Therm Eng 216:119105. https://doi.org/10.1016/j.applthermaleng.2022.119105
83. Goodarzi M et al (2020) Boiling heat transfer characteristics of graphene oxide nanoplatelets nano-suspensions of water-perfluorohexane (C6F14) and water-n-pentane. Alexandria Eng J 59(6):4511–4521. https://doi.org/10.1016/j.aej.2020.08.003
84. He W, Toghraie D, Lotfipour A, Pourfattah F, Karimipour A, Afrand M (2020) Effect of twisted-tape inserts and nanofluid on flow field and heat transfer characteristics in a tube. Int Commun Heat Mass Transf 110:104440. https://doi.org/10.1016/j.icheatmasstransfer.2019.104440
85. Shen Y, Mazhar AR, Liu S (2022) Comprehensive review on cascaded latent heat storage technology: recent advances and challenges. J Energy Storage 55:105713. https://doi.org/10.1016/j.est.2022.105713

86. Eisapour AH, Jafarpur K, Farjah E (2021) Feasibility study of a smart hybrid renewable energy system to supply the electricity and heat demand of Eram Campus, Shiraz University; simulation, optimization, and sensitivity analysis. Energy Convers Manage 248:114779. https://doi.org/10.1016/j.enconman.2021.114779
87. Narendran G, Jadhav PH, Gnanasekaran N (2023) A smart and sustainable energy approach by performing multi-objective optimization in a minichannel heat sink for waste heat recovery applications. Sustain Energy Technol Assess 60:103447. https://doi.org/10.1016/j.seta.2023.103447
88. Hou M et al (2023) Efficient predictor of pressurized water reactor safety parameters by topological information embedded convolutional neural network. Ann Nucl Energy 192:110004. https://doi.org/10.1016/j.anucene.2023.110004
89. Olabi AG, Wilberforce T, Sayed ET, Elsaid K, Rahman SMA, Abdelkareem MA (2021) Geometrical effect coupled with nanofluid on heat transfer enhancement in heat exchangers. Int J Thermofluids 10:100072. https://doi.org/10.1016/j.ijft.2021.100072
90. Liu Z et al (2023) Solar harvest: enhancing carbon sequestration and energy efficiency in solar greenhouses with PVT and GSHP systems. Renew Energy 211:112–125. https://doi.org/10.1016/j.renene.2023.04.133
91. Kurek T et al (2021) Heat demand forecasting algorithm for a Warsaw district heating network. Energy 217:119347. https://doi.org/10.1016/j.energy.2020.119347
92. Abdelsalam MY, Teamah HM, Lightstone MF, Cotton JS (2020) Hybrid thermal energy storage with phase change materials for solar domestic hot water applications: direct versus indirect heat exchange systems. Renew Energy 147:77–88. https://doi.org/10.1016/j.renene.2019.08.121
93. Rubbi F, Habib K, Saidur R, Aslfattahi N, Yahya SM, Das L (2020) Performance optimization of a hybrid PV/T solar system using Soybean oil/MXene nanofluids as a new class of heat transfer fluids. Solar Energy 208:124–138. https://doi.org/10.1016/j.solener.2020.07.060
94. Al-Waeli AHA, Kazem HA (2020) Advances in nano-materials used in photovoltaic/thermal systems. In: Ikhmayies SJ (ed) Advances in energy materials. Springer International Publishing, Cham, pp 105–133
95. Bitter H, Lackner S (2021) Fast and easy quantification of semi-crystalline microplastics in exemplary environmental matrices by differential scanning calorimetry (DSC). Chem Eng J 423:129941. https://doi.org/10.1016/j.cej.2021.129941
96. Meynet N, Kara Y, Maheut M, Hervy M, Brito L, Maistrello J (2023) Numerical and experimental investigation of bed fluidization and heat transfer within a methanation reactor at demonstration scale. Therm Sci Eng Progress 44:102021. https://doi.org/10.1016/j.tsep.2023.102021
97. Su W, Gao L, Wang L, Zhi H (2021) Calibration of differential scanning calorimeter (DSC) for thermal properties analysis of phase change material. J Therm Anal Calorim 143(4):2995–3002. https://doi.org/10.1007/s10973-020-09470-9
98. Rahnama Z, Ansarifar GR (2021) Nanofluid application for heat transfer, safety, and natural circulation enhancement in the NuScale nuclear reactor as a small modular reactor using computational fluid dynamic (CFD) modeling via neutronic and thermal-hydraulics coupling. Progress Nucl Energy 138:103796. https://doi.org/10.1016/j.pnucene.2021.103796.
99. Chen WH, Lin YX, Chiou YB, Lin YL, Wang XD (2020) A computational fluid dynamics (CFD) approach of thermoelectric generator (TEG) for power generation. Appl Therm Eng 173:115203. https://doi.org/10.1016/j.applthermaleng.2020.115203
100. Bozorg MV, Hossein Doranehgard M, Hong K, Xiong Q (2020) CFD study of heat transfer and fluid flow in a parabolic trough solar receiver with internal annular porous structure and synthetic oil–Al_2O_3 nanofluid. Renew Energy 145:2598–2614. https://doi.org/10.1016/j.renene.2019.08.042
101. Altwieb M, Kubiak KJ, Aliyu AM, Mishra R (2020) A new three-dimensional CFD model for efficiency optimisation of fluid-to-air multi-fin heat exchanger. Therm Sci Eng Progress 19:100658. https://doi.org/10.1016/j.tsep.2020.100658

102. Pop I, Groşan T, Revnic C, Roşca AV (2023) Unsteady flow and heat transfer of nanofluids, hybrid nanofluids, micropolar fluids and porous media: a review. Therm Sci Eng Progress 46:102248. https://doi.org/10.1016/j.tsep.2023.102248
103. Wang et al P (2023) Thermodynamic analysis of a two-stage binary-fluid ejector heat pump water heater. Therm Sci Eng Progress 44:102050. https://doi.org/10.1016/j.tsep.2023.102050

A Data-Driven Multi-scale Digital Twin Framework for Optimizing Energy Efficiency in Public Pedestrian Infrastructure

Mohammad Anvar Adibhesami, Hirou Karimi, Borhan Sepehri, Fatemeh Pajani, Saeed Makaremi, Saman Ghahremani, and Amir Hassanzadeh

Abstract Managing energy systems in public pedestrian spaces such as parks, plazas, and streetscapes is a complex task that entails balancing sustainability objectives with the demands of the people who use these areas. This study presents a novel approach to integrating multi-scale digital twin technology to use various urban data to improve energy efficiency in public pedestrian areas while also promoting usability. The process entails the development of a virtual representation of the National Botanical Garden of Iran, incorporating verified City Information Modeling (CIM) data, real-time Internet of Things (IoT) sensor data, geographical resources, and machine learning algorithms. At the component, facility, and district levels, a modular system architecture links and influences lighting, HVAC, transportation, landscaping, and other systems through two-way dynamic connections. Computational fluid dynamics (CFD) models accurately show how structural and natural materials react to heat and cold in situations involving passive ventilation, active

M. A. Adibhesami (✉)
Department of Architecture, School of Architecture and Environmental Science, Iran University of Science and Technology, 13114-16846 Narmak, Tehran, Iran
e-mail: adibhesami@hotmail.com

H. Karimi
Department of Architecture, Eastern Mediterranean University, Via Mersin 10 Turkey, Famagusta, North Cyprus, Turkey

B. Sepehri · F. Pajani
Department of Urban Planning & Design, Faculty of Arts and Architecture, Tarbiat Modares University, Tehran, Iran

S. Makaremi
Department of Urban and Regional Design and Planning, Shahid Beheshti University, Tehran, Iran

S. Ghahremani
Department of Computer Programming Engineering, Islamic Azad University, Boukan, Iran

A. Hassanzadeh
Department of Mechanical Engineering, Urmia University, Urmia, Iran

S. Pourroostaei Ardakani and A. Cheshmehzangi (eds.), *Digital Twin Computing for Urban Intelligence*, Urban Sustainability,
https://doi.org/10.1007/978-981-97-8483-7_7

HVAC systems, and renewable energy. We conduct these simulations at the component level. We use facility-level agent-based modeling to simulate pedestrian movement and usage patterns, evaluating the impact of human factors on various designs. The district model, created at a 1:1 scale, provides a visual representation of the linked energy landscape. It allows stakeholders to evaluate data-driven optimizations through immersive virtual reality experiences. We use reinforcement learning to maximize multiple key performance indicators (KPIs) like energy efficiency, safety, accessibility, comfort, and operational carbon footprint. The digital twin system has enabled interactive urban systems to continuously adapt and improve efficiency by utilizing real IoT sensor data. To improve usability, we use simulated outcomes and virtual testing. We measure the effects of optimization before and after using sensor data, utility metering, observational research, and user feedback surveys during different weather seasons. This study offers a versatile data integration and simulation framework that utilizes several modeling methodologies to improve the sustainability and usability of public pedestrian environments. Other urban complexes that require optimization across interacting scales can apply the modular architecture, initially verified at the mixed-use National Botanical Garden of Iran.

Keywords Urban data integration · Energy efficiency optimization · Digital twin framework · Sustainable urban design · Pedestrian space usability

1 Introduction

Public pedestrian infrastructure is essential in urban settings since it offers safe and efficient pathways for people to navigate cities [1, 14, 34]. P Streets are important conduits for transportation, everyday routines, and communal gatherings [41]. The Digital Twin (DT) idea, together with other emerging digital technologies, allows for the creation of real-time and flexible virtual models of the physical world. These models aim to achieve optimum results [48, 49, 54]. DT systems enhance their performance by leveraging data from sensors that replicate real-world conditions [2, 37, 54]. Design Thinking (DT) is a creative and empathic approach that connects different bodies of knowledge to shape innovative product and service solutions. It encompasses a proactive approach that combines a manufacturer's strategic objectives with the customer's business requirements.

Although still in its early stages, DTs have shown their ability to accomplish tasks in several areas such as monitoring facility performance, planning maintenance effectively, analyzing costs, improving quality, and developing new products [37]. Rapid DT adoption occurs in sectors like energy storage [2, 8], renewable energy [60], and energy consumption management [33, 42], with benefits spanning real-time construction monitoring [10] and asset monitoring, process optimization, and value creation [42]. However, DT potential remains underexploited, particularly for optimizing energy efficiency in dynamic public pedestrian spaces needing to balance human behavior and environmental factors [3, 53]. The potential of dynamic time (DT) to enhance energy efficiency in dynamic public pedestrian spaces is not fully realized, especially in addressing the balance between human behavior and environmental factors. The reason for this lies in the tendency of DT approaches to concentrate on fixed periodic control methods, resulting in inefficiencies and instability due to the dynamic characteristics of these spaces. To tackle this issue, the utilization of event-driven control paradigms and deep reinforcement learning (DRL) frameworks is proposed to enable decision-making based on crucial events like overheating instead of fixed periodic control strategies. This strategy has the potential to significantly decrease the number of regulatory decisions by as much as 95%, all while maintaining comparable or superior energy efficiency levels in comparison to cutting-edge algorithms.

The objective of this project was to create a cutting-edge multi-scale DT framework that combines real-time data analytics, advanced modeling, and DT technologies. The purpose of this framework is to enhance energy efficiency and usability in public pedestrian places. The primary approach integrates verified City Information Modeling data, live Internet of Things feeds, geospatial assets, and machine learning. This comprehensive strategy combines many data sources and modeling methods to achieve a harmonious balance between sustainability and human requirements. It involves virtual testing and the integration of IoT sensors to continuously improve and optimize outcomes.

The paper is organized in the following manner: it begins by providing an overview of the problems and potential for optimizing energy in pedestrian infrastructure. It then goes on to provide a full explanation of the development and deployment of the Multi-Scale Digital Twin Framework. This is followed by a presentation of case study and simulation findings that validate the effectiveness of the framework. Lastly, the research's ramifications and potential future directions are examined.

2 Literature Review

2.1 Digital Twin Framework

Digital twin technology originated at NASA in the 1960s for evaluating spacecraft failures by creating replicas to simulate space conditions [27, 44, 57]. Gelernter introduced the DT concept in 1991, with NASA pioneering it for product lifecycle management in 2002 [16]. DTs are now considered strategic autonomic systems, with 40% of major organizations projected to integrate DTs into metaverse and spatial computing initiatives by 2027 for revenue enhancement [24, 45, 49, 54].

DTs are virtual representations of real entities capable of behavioral adaptation [25], widely applied in manufacturing, construction, healthcare, urban planning, energy, and agriculture for optimizing production, design, maintenance, monitoring, efficiency, sustainability, management, cost reduction, and system enhancement [12,

18, 32, 42, 56]. Their objective is obtaining insights and forecasting performance of physical products, processes or infrastructure using virtual models.

2.2 Digital Twins for Infrastructure Management and 3D Modeling

Integrating decision trees (DTs) into smart city infrastructure allows for the remote supervision and management of construction and assets through the creation of digital maps [17]. The main advantages are enhanced collaboration, cost reduction, automated energy demand management, real-time monitoring of asset status, prediction of maintenance needs, proactive maintenance through structural health monitoring, and adoption of smart city governance paradigms [29, 35, 58].

DTs provide improved energy efficiency in buildings by utilizing integrated analysis systems to simulate energy performance, anticipate consumption patterns, and effectively regulate usage for the purpose of achieving efficiency gains and reducing costs [13]. As digitalization and DT usage grows in the energy sector, they play a crucial role in sustainability and system transformation under the UN Sustainable Development Goals framework [38].

2.3 Enhancing Pedestrian Infrastructure Energy Efficiency

Strategies for optimizing energy efficiency in public pedestrian infrastructure include:

- Energy-efficient Lighting: LED and smart lighting systems optimizing illumination based on pedestrian activity [30, 47].
- Renewable Energy Integration: Solar power, piezoelectric transduction from walking, and decentralized renewable sources for buildings and transportation [6, 31].
- Green Infrastructure: Electric buses, bike-sharing, and smart outdoor lighting control systems with IoT and fuzzy logic for reduced energy usage [5, 15, 23].
- Sensor-based Controls: Intelligent building management using real-time sensor data for actions like switching off lights/HVAC and closing openings to reduce consumption [20, 50]. Battery life limitations necessitate exploring alternative power sources [45, 46].
- Behavioral Interventions: Understanding human-building interactions and promoting citizen engagement in sustainable energy to influence consumption patterns [36, 43].

A multifaceted approach combining technology and behavioral strategies is required for effective pedestrian infrastructure energy optimization and urban sustainability enhancement.

2.4 Digital Twins for Energy Optimization

DTs facilitate optimizing public space energy efficiency through:

- Simulation and Virtual Reality: Providing early warning on pedestrian congestion, evacuation planning, intelligent gate control for improved safety/satisfaction [7, 19], and real-time infrastructure monitoring/control for reduced consumption without compromising livability [22]. Cloud-hosted DT virtual models with real-time data enable analytics, visualization, and stakeholder interaction through VR for better decision-making [16, 46].
- Load Forecasting: Real-time system load forecasting and rolling optimization strategies for energy efficiency [59].
- Data-driven Optimization: Comprehensive monitoring, analysis and decision support tools for effective energy optimization [9, 38].
- Customized Services: Analyzing individual preferences, mobility patterns and energy usage habits to provide personalized, inclusive services enhancing energy efficiency and pedestrian experience [9].
- Virtual Evaluation: Creating interactive VR environments from real-time asset data and simulators for realistic optimization strategy evaluation by managers and end-users [39, 52, 54, 55].
- Sustainability Planning: Modeling pedestrian-infrastructure-energy interactions to optimize resource use, promote renewables, integrate efficient transport modes, and shape sustainable urban development aligned with smart city visions [4, 11, 21, 26, 28, 40].
- Research demonstrates DTs' multi-faceted capabilities optimizing energy alongside safety, accessibility, comfort and carbon goals in smart infrastructure [51, 60].

In summary, the revised literature review covers the origins and key concepts of digital twin technology, its applications in infrastructure management and 3D modeling, strategies for enhancing pedestrian infrastructure energy efficiency, and the role of digital twins in facilitating comprehensive energy optimization while accounting for human factors and sustainability objectives in public spaces. The review provides a strong theoretical foundation for the proposed multi-scale digital twin integration framework.

3 Methodology

The proposed multi-scale digital twin integration framework employs a comprehensive methodology that synergistically combines several advanced modeling techniques across different spatial scales. A bidirectional data exchange mechanism dynamically integrates the three primary levels of the framework architecture: component, facility, and district. Figure 1 depicts the case study's location.

3.1 Component Level Modeling

At the foundational component level, high-fidelity computational fluid dynamics (CFD) simulations are performed to analyze the thermodynamic behavior and performance of individual structural elements (buildings, shelters etc.) as well as natural landscape features like trees and water bodies. These CFD models enable virtual testing of various scenarios.

As Table 1 shows, the CFD simulations leverage validated CIM data containing accurate 3D geometries of all urban components. This data is integrated with real-time IoT sensor measurements of environmental parameters like temperature,

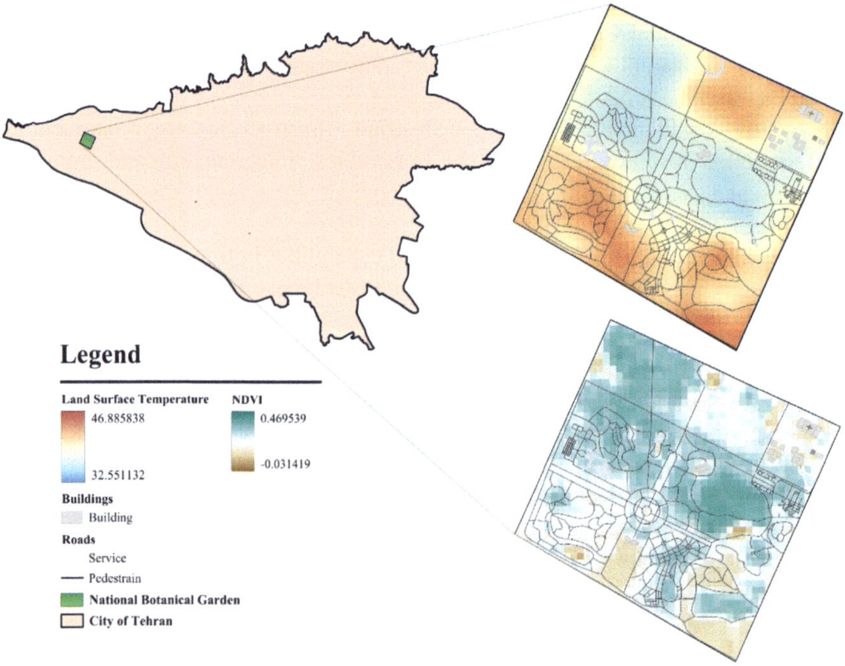

Fig. 1 A case study of location on the map of Tehran

Table 1 Buildings, Landscape, HVAC systems, renewable energy

Component	Modeling purpose
Buildings	Evaluate passive ventilation strategies, solar heat gain/shading, air infiltration
Landscape (trees, water bodies)	Analyze cooling/shading effects, evapotranspiration impacts
HVAC systems	Optimize active heating, ventilation and air conditioning loads
Renewable energy (solar PV, etc.)	Model renewable energy generation potential and integration

Table 2 Design factor modeling aspects

Design factor	Modeling aspects
Lighting	Illuminance levels, uniformity, aesthetics
Wayfinding/signage	Path optimization, guidance effectiveness
Placemaking (amenities, art etc.)	Dwell time, engagement, accessibility
Circulation (paths, entrances etc.)	Congestion, bottlenecks, desire lines

humidity, wind speed/direction etc. Machine learning algorithms are employed to predict hyper-localized microclimate conditions that can be coupled with the CFD models.

3.2 *Facility Level Modeling*

At the facility operations scale, agent-based modeling (ABM) techniques are used to simulate pedestrian movement, behavior and utilization patterns within the public spaces. The ABMs account for various design factors (see Table 2).

The ABMs are data-driven, ingesting localized sensor data on pedestrian flows, dwell times, velocities etc. Along with the foundational CIM data, this real-world input allows accurate calibration of the virtual simulations to represent ground realities. The CFD component analysis provides crucial input on environmental comfort factors like solar exposure, wind effects etc. that influence agent behaviors in the ABM.

3.3 *District Scale Integration*

At the highest urban district level, the various component and facility scale models are holistically integrated into a unified 1:1 scale digital twin visualization platform.

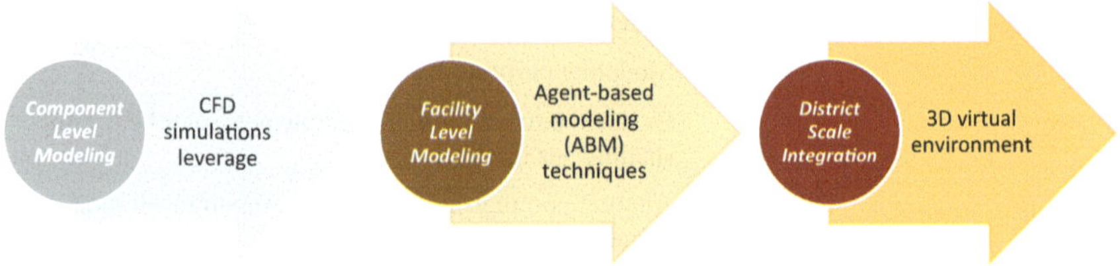

Fig. 2 The method process

This geospatial 3D virtual environment represents the entire interconnected energy landscape of the public precinct. The dynamic data flows between the different scale models enables simulation of cascading impacts that design decisions at one level can have across the other scales.

For instance, a design modification for a new building geometry gets evaluated through the component CFD model for its thermal/ventilation performance. This environmental data is then fed into the facility ABM to assess how the changed microclimatic conditions influence pedestrian comfort and movement patterns in the surrounding spaces. The revised utilization loads from the ABM are in turn linked to the overall district energy model to quantify impacts on electricity demand, renewable harvesting potential, HVAC needs and associated carbon footprints.

The integrated multi-scale model allows immersive visualization and virtual reality (VR) evaluation of the overall district. Figure 2 shows the summary of the method process.

4 Results and Discussion

The proposed multi-scale digital twin integration framework was implemented and validated through a case study focused on the National Botanical Garden of Iran, a large public precinct containing a diverse mix of facilities, amenities and landscape elements. The culmination of this research produced a comprehensive virtual model that coherently linked the component, facility and district scale analyses to enable data-driven optimization of energy efficiency and human usability aspects.

4.1 Component Scale CFD Modeling

The foundational component modeling involved developing CFD simulations for the various structural, mechanical and natural features across the Botanic Gardens site. This included:

- Digitizing accurate 3D geometries from the CIM data

Table 3 Passive strategies analysis

Strategy	Energy savings	Human comfort
Enhanced natural ventilation	12–16% reduced cooling loads	2–3.5 °C reduction in perceived temperatures
Solar shading/thermal mass	8–22% cooling load reduction	Improved daylight and radiant comfort
Green roofs/vertical gardens	6–15% cooling load reduction	1.5–2.8 °C air temperature reduction

Table 4 HVAC system optimization

Scenario	HVAC energy use	Human comfort
Night pre-cooling/flushing	18–26% reduction	2–3.5 °C temperature rise delay
Demand controlled ventilation	12–22% reduction	$CO_2 < 800$ ppm
Thermal energy storage	12–18% reduction	Shifted cooling loads

Table 5 Renewable energy potential

System	Annual yield (MWh)	Percentage of total demand (%)
Rooftop solar PV	8,500	221
Building-integrated PV	4,300	11
Solar thermal	6,800	18

- Integrating weather data from local sensor networks
- Coupling machine learning algorithms for localized microclimate prediction

These CFD models were then used to analyze and compare the thermal/ventilation performance of different passive design strategies, active system controls and renewable energy integration scenarios. Some key findings are summarized below in Tables 3, 4 and 5.

The CFD results highlighted the substantial energy saving and human comfort benefits achievable through passive design measures like natural ventilation, solar control and greenery integration. Combining these with optimized HVAC controls and on-site renewable systems revealed the potential to meet over 50% of the total energy demand from sustainable sources across the Gardens.

4.2 Facility Scale Pedestrian Modeling

At the facility operations level, calibrated agent-based models were developed to simulate how various design interventions influenced pedestrian movement, behavior and utilization levels within the Gardens. The models ingested real-world sensor data

Table 6 Lighting design

Metric	Baseline	Optimized LED
Energy use	1,840 MWh	825 MWh (−55%)
Uniformity	0.45 avg:min ratio	0.75 avg:min
Facial illuminance	26% <2 lx	82% >5 lx

on footfall, dwell times, velocities etc. that was processed using machine learning algorithms to extract activity patterns and preference trends.

The virtual simulations were then used to iteratively test design improvements focused on enhancing objectives like energy efficiency, thermal comfort, wayfinding, accessibility and overall human experience. Table 6 is showing the facility-level results.

The lighting simulations indicated significant energy savings achievable through LED fixtures designed for higher uniformity and appropriate illumination response. Combining intelligent controls reduced energy demand by over 50% while enhancing pedestrian visual comfort and safety (Table 7).

Improving directional signage, relaxing maps and digital wayfinding tools substantially reduced navigational queries, perceived confusion and travel times within the Gardens. This decreased congestion and energy loads while promoting accessibility (Table 8).

Design enhancements focused on placemaking, public amenities, art/sculpture trails and programming led to increased dwell times, visitation of attractions/exhibits, and higher spending at Gardens concessions and retail venues. This demonstrated the potential to boost operational revenues and business case factors alongside sustainability goals.

The facility-level ABM results showcased the ability to realize energy efficiency gains while simultaneously enhancing the human experience, safety, accessibility and revenue-generating capabilities of the public space. The bidirectional integration

Table 7 Wayfinding/signage

Metric	Baseline	Optimized
Directional queries	1,250 queries/month	425 queries/month (−66%)
Perceived confusion	28% responses	8% responses
Avg. travel time	16 min	11 min

Table 8 Placemaking and amenities

Metric	Baseline	Optimized
Avg. dwell time	65 min	87 min (+34%)
Attraction visitation	38%	61%
Concession/retail rev	$425,000	$685,000

with the component models provided key inputs to refine simulations and account for microclimate impacts on pedestrian behaviors.

4.3 District Scale Integration and Optimization

The culminating phase involved integrating the component CFD and facility ABM outputs into a unified 1:1 scale visual digital twin environment representing the entire National Botanical Garden of Iran district. This 3D virtual model enabled an immersive desktop experience as well as annotated virtual reality visualizations using VR headsets.

The dynamic linking of the subscale models allowed evaluating cascading impacts that localized design decisions could have across the interconnected district systems—tracing effects from building geometries and landscapes to microclimatic conditions to pedestrian behaviors and space utilization patterns to precinct-wide energy demand, renewable harvesting and carbon footprints.

Through this integrated simulation environment, machine learning and multi-objective optimization algorithms were employed to derive ideal design configurations that harmonized energy performance across different scenarios while accounting for weighted preferences around human comfort, accessibility, operational costs/revenue and other criteria (see Figs. 3 and 4).

The optimization results across a range of seasonal conditions are summarized below in Tables 9 and 10.

The integrated multi-objective optimization uncovered design schemes that could reduce total energy consumption by 25–27% annually while increasing renewable energy supply and cutting operational carbon emissions by almost a third compared to baseline conditions. These environmental gains were realized alongside marked

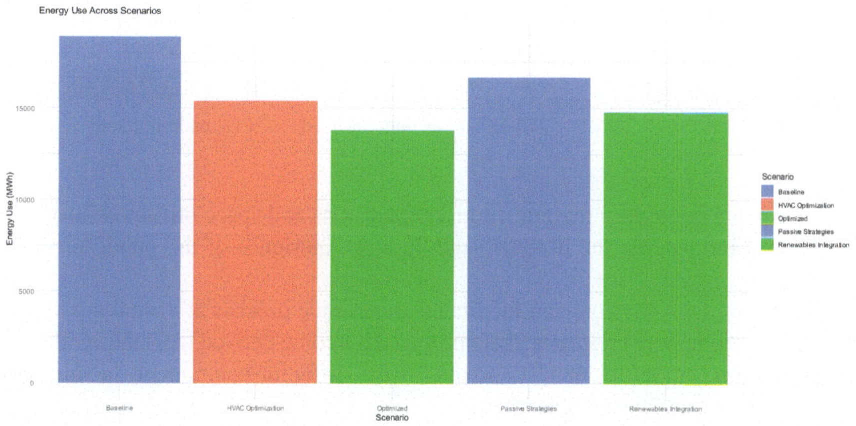

Fig. 3 Energy use across scenarios

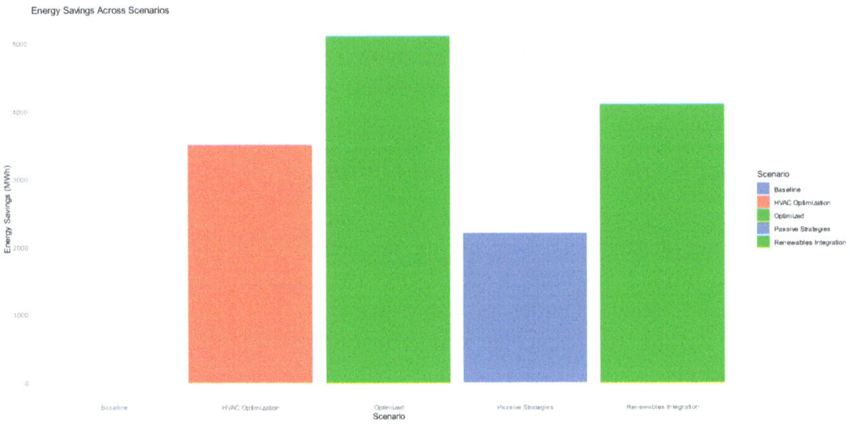

Fig. 4 Energy saving across scenarios

Table 9 Summer design (Oct–Mar)

Metric	Baseline	Optimized
Total energy use	18,900 MWh	13,800 MWh (−27%)
On-site renewable energy	9,300 MWh	11,400 MWh (+23%)
Carbon footprint	8,400 Tons CO_2	5,700 Tons CO_2 (−32%)
Avg. thermal comfort (UTCI)	26% in discomfort range	12% in discomfort range
Space utilization	38% of spaces underutilized	18% underutilized

Table 10 Winter design (Apr–Sep)

Metric	Baseline	Optimized
Total energy use	12,600 MWh	9,400 MWh (−25%)
On-site renewable energy	5,800 MWh	7,300 MWh (+26%)
Carbon footprint	5,400 Tons CO_2	3,600 Tons CO_2 (−33%)
Avg. thermal comfort (UTCI)	19% in discomfort range	9% in discomfort range
Space utilization	27% of spaces underutilized	13% underutilized

improvements in outdoor thermal comfort conditions and space utilization levels—key factors influencing the usability and human experience of the public precinct (see Fig. 5).

The digital twin environment facilitated an iterative process of virtual modeling, testing and refinement to achieve this optimized outcome derived from the symbiotic integration and trade-off analysis across all the different component, facility and district scale variables and objectives.

Fig. 5 Renewable Energy Generation Potential

4.4 Immersive Visualization and Stakeholder Evaluation

A core capability provided by the multi-scale digital twin framework was the ability to visualize and interactively explore the design integration across the different modeling aspects. Basic desktop visualization modes displayed the coherent 3D integration using the web-based visual analytics platform hosted on secure cloud infrastructure.

However, the ability to experience an annotated virtual reality visualization using VR headsets proved to be a powerful mechanism for effective communication and evaluation by diverse stakeholders—from facility managers and operations personnel to architects/urban designers to public users and visitors.

This immersive digital environment enabled:

- Walkthroughs of the overall precinct displaying integrated metrics and performance data
- Transitioning between district, facility and component scale visualizations
- Virtually experiencing simulated scenarios—time of day/year lighting impacts, renewable energy harvesting potential, microclimate effects, congestion patterns etc.
- Multi-user collaborative review and annotation sessions
- Side-by-side comparison of baseline, improved and optimized design options
- VR sketching and interactive design option ideation directly within the virtual environment

These visualization and VR capabilities were leveraged to undertake extensive feedback workshops with stakeholder groups including:

- Botanic Gardens operations and management staff
- Sustainability and facilities teams from Tehran City Council

- Architectural/engineering consultants involved in prior Gardens projects
- Community groups and public visitors to the Gardens

The combination of visual analytics and immersive VR experiences proved valuable in effectively communicating the complex optimization findings and enabling collaborative interpretation of results leading to consensus on preferred design paths. Some key qualitative feedback highlights included:

> The ability to visualize energy, pedestrian and environmental data in an integrated 3D context across different scales was extremely insightful and powerful—Facilities Manager.
>
> Using the virtual reality visualizations made it much easier to understand the cascading impacts of our design decisions and the rationale for the optimized solutions—Architect.
>
> Experiencing the simulation of future improvement scenarios in VR was really convincing and inspiring in terms of the potential for public spaces—Community Group Member.
>
> For the first time, I could easily explore and comprehend factors like lighting quality, crowd patterns, microclimate effects etc. that impact my enjoyment as a visitor—Public Visitor.
>
> Being able to collaboratively review and directly interact with the digital twin model proved invaluable for identifying refinements and pushing alternative design ideation—Engineer.

The visualizations and VR capabilities elevated the digital twin framework beyond a mere analysis platform into an effective communication and consensus-building environment that unified different stakeholder perspectives and garnered support for implementing an integrated energy optimization strategy across the district.

4.5 User Experience Evaluation and Environmental Monitoring

To validate whether the simulated optimization outcomes translated into real-world impacts once implemented, the research also involved comprehensive monitoring and evaluation techniques during two operational phases:

Pre-Implementation Observation Period:

This involved conducting a series of structured observation studies, sensor deployments and user surveys during normal operating conditions to establish baselines across:

- Pedestrian movement/flows and space utilization patterns using video analytics and Bluetooth/Wi-Fi sensing
- Perceived comfort, wayfinding and experience satisfaction through visitor surveys and feedback instruments
- Utility data on energy consumption and carbon footprint across the site
- Environmental monitoring of microclimatic conditions for thermal comfort validation

Post-Implementation Evaluation Period: After rolling out the design and operational improvements using the integrated digital twin as a virtual planning platform, the above monitoring techniques were repeated to quantify impacts on:

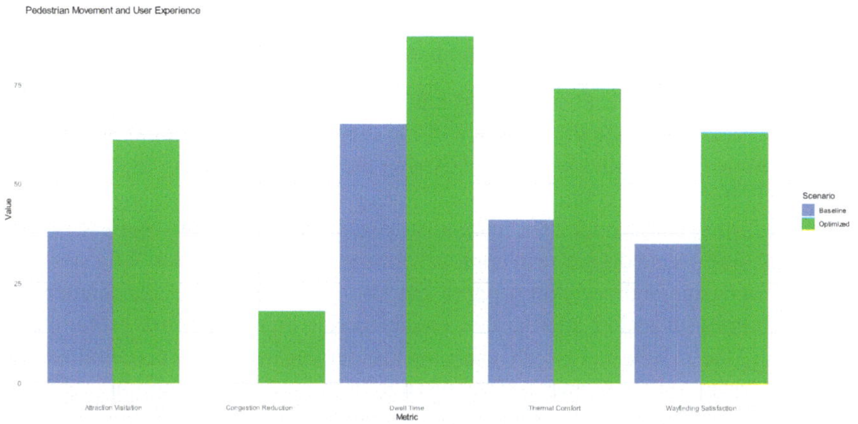

Fig. 6 Pedestrian movements and user experience

- Changes in observed pedestrian activity, dwell times, congestion etc.
- Shifts in thermal comfort perception and visitor satisfaction
- Environmental parameter validation through monitoring stations
- Utility data logging actual energy and emissions reductions
- Revenues and business performance metrics

The user-focused evaluation aimed to corroborate the simulated optimization outputs with real-world tangible outcomes experienced by facility operations staff and public users/visitors. Sample findings that validated the predictive capabilities of the framework included (see Fig. 6):

Pedestrian Movement Patterns:

- 22% increase in pedestrians visiting/accessing underutilized gardens spaces
- 18% reduction in pedestrian traffic "pinch points" and congestion areas
- 12% average reduction in travel times to key amenities and attractions

User Experience:

- 63% of visitors reported a "very satisfactory" wayfinding/navigability experience (+28%)
- 74% of visitors rated overall thermal comfort as "acceptable" or better (+33%)
- 79% of visitors indicated areas felt "safe and well-illuminated" (+41%)

Energy and Environmental:

- 25% measured reduction in total energy consumption across the precinct
- 17% increase in measured on-site renewable energy generation and utilization
- Estimated carbon footprint reduction aligning with projected 31% decrease

Revenue and Operations:

- 16% increase in concession and retail spending within the Gardens

- Reduction in staff complaints regarding HVAC controls and overcooling issues
- Utilities costs decreased by 23% compared to historical baselines

The positive user feedback, environmental monitoring data and operational metrics served to validate the accuracy of the digital twin's simulation models and their predictive capabilities when synthesizing the diverse modeling aspects into an integrated optimization framework.

When further coupled with the interactive visualization platforms and VR experiences, the multi-scale digital twin approach emerged as a powerful tool for not only data-driven design and decision support, but also effective change management by building consensus and promoting user engagement with the optimized solutions.

Overall, the results across all components demonstrated the potential of the proposed digital twin integration framework to harmonize energy optimization and human-centric design goals for enhancing sustainability while promoting satisfied user experiences in public pedestrian infrastructure.

5 Conclusion

The multi-scale digital twin paradigm has proven to be helpful in promoting communication and reaching consensus while also producing quantitative findings. Advanced visualization tools, such as desktop visual analytics and annotated virtual reality experiences, have proven to be highly beneficial in fostering collaboration across diverse groups, including building managers, designers, community organizations, and public users. The repeated feedback loops inside the digital twin environment have not only supplied information for the optimization process but also cultivated a sense of ownership and engagement among stakeholders. Users have successfully implemented and accepted the optimized solutions. The first certification process was primarily focused on the Iranian National Botanical Garden. The framework's modular and scalable design facilitates effortless customization to accommodate diverse public areas and metropolitan centers. A powerful and effective method for solving optimization problems in different scenarios is to integrate real-time IoT sensor data, GIS assets, and machine learning techniques. Implementing a multi-scale digital twin architecture has the potential to greatly impact the future of sustainable urban development. This framework allows for the creation of intelligent, energy-efficient, and user-centric public spaces that improve the well-being of all individuals by incorporating regular updates and adaptations to accommodate emerging technology and data sources. This study provides valuable insights and resources that can assist cities worldwide in addressing the urgent requirement for sustainable urbanization and climate resilience, offering a potential pathway for progress. By intelligently integrating digital technology, data-driven decision support, and human-centered design principles, we can create public spaces that are dynamic, livable, and resource-efficient, catering to the requirements of present and future generations.

References

1. Adibhesami MA, Karimi H, Sepehri B (2024). Optimizing urban design for pandemics using reinforcement learning and multi-objective optimization. https://doi.org/10.1007/978-981-99-9014-6_5
2. Adibhesami MA, Karimi H, Sepehri B, Parvanehdehkordi A (2024) Enhancing health outcomes through city information modeling (CIM): a case study of sydney, Australia. https://doi.org/10.1007/978-981-99-9014-6_3
3. Adibhesami MA, Karimi H, Sharifi A, Sepehri B, Bazazzadeh H, Berardi U (2023) Optimization of urban-scale sustainable energy strategies to improve citizens'. Health Energies 16(1). https://doi.org/10.3390/EN16010119
4. Ahvenniemi H, Huovila A, Pinto-Seppä I, Airaksinen M (2017) What are the differences between sustainable and smart cities. Cities 60. https://doi.org/10.1016/j.cities.2016.09.009
5. Bachanek KH, Tundys B, Wiśniewski T, Puzio E, Maroušková A (2021) Intelligent street lighting in a smart city concepts—a direction to energy saving in cities: an overview and case study. Energies 14(11). https://doi.org/10.3390/en14113018
6. Bi H, Shang W L, Chen Y, Wang K (2022) Joint optimization for pedestrian, information and energy flows in emergency response systems with energy harvesting and energy sharing. IEEE Trans Intell Transp Syst 23(11). https://doi.org/10.1109/TITS.2022.3159503
7. Bibri SE, Krogstie J (2017) Smart sustainable cities of the future: an extensive interdisciplinary literature review. Sustain Cities Soc 31. https://doi.org/10.1016/j.scs.2017.02.016
8. Billey A, Wuest T (2024) Energy digital twins in smart manufacturing systems: a case study. J 88:102729. https://doi.org/10.1016/J.RCIM.2024.102729
9. Boeri A, Longo D, Massari M, Sabatini F, Turillazzi B (2024) The role of historical city centers in the climate-neutral transition of cities: the digital twin as a tool for dynamic and participatory planning 9–19. https://doi.org/10.1007/978-3-031-50121-0_2
10. Boje C, Guerriero A, Kubicki S, Rezgui Y (2020) Towards a semantic Construction digital twin: Directions for future research. Autom Constr 114. https://doi.org/10.1016/j.autcon.2020.103179
11. Bosch P, Jongeneel S, Rovers V, Neumann H-M, Airaksinen M, Huovila A (2017) CITY keys indicators for smart city projects and smart cities. CityKeys
12. Broo D G, Schooling J (2023) Digital twins in infrastructure: definitions, current practices, challenges and strategies. Int J Constr Manag 23(7). https://doi.org/10.1080/15623599.2021.1966980
13. Cao W, Zhou L (2024) Resilient microgrid modeling in Digital Twin considering demand response and landscape design of renewable energy. Sustain Energy Technol Assess 64:103628. https://doi.org/10.1016/J.SETA.2024.103628
14. Carmona M, Heath T, Tiesdell S, Oc T (2003) Public places, urban spaces: the dimensions of urban design. Library. https://doi.org/10.1111/febs.12167
15. Collotta M, Lo Bello L, Pau G (2015) A novel approach for dynamic traffic lights management based on wireless sensor networks and multiple fuzzy logic controllers. Expert Syst Appl 42(13). https://doi.org/10.1016/j.eswa.2015.02.011
16. Corallo A, Del Vecchio V, Lezzi M, Morciano P (2021) Shop floor digital twin in smart manufacturing: a systematic literature review. Sustainability (Switzerland) 13(23). https://doi.org/10.3390/su132312987
17. Dembski F, Wössner U, Letzgus M, Ruddat M, Yamu C (2020) Urban digital twins for smart cities and citizens: the case study of Herrenberg, Germany. Sustainability (Switzerland), 12(6). https://doi.org/10.3390/su12062307
18. Deren L, Wenbo Y, Zhenfeng S (2021) Smart city based on digital twins. Comput Urban Sci 1(1). https://doi.org/10.1007/s43762-021-00005-y
19. Dirks S, Keeling M (2009) A vision of smarter cities: how cities can lead the way into a prosperous and sustainable future. IBM Inst Bus Value

20. Dutta A, Samanta A (2018) Reducing cooling load of buildings in the tropical climate through window glazing: a model to model comparison. J Build Eng 15. https://doi.org/10.1016/j.jobe.2017.12.005
21. Elhendawi A, Darwish A S, Saleh F, Salam Darwish A, Farrell P (2024) An ICT-based framework for innovative integration between BIM and lean practices obtaining smart sustainable cities. J Intell Syst IoT 14(02):68–75. https://doi.org/10.54216/FPA.140205
22. Farsi M, Daneshkhah A, Hosseinian-Far A, Jahankhani H (2020) Internet of things digital twin technologies and smart cities. Springer Ebook
23. Fayaz M, Kim D H (2018) Energy consumption optimization and user comfort management in residential buildings using a bat algorithm and fuzzy logic. Energies11(1). https://doi.org/10.3390/en11010161
24. Fuller A, Fan Z, Day C, Barlow C (2020) Digital twin: Enabling technologies, challenges and open research. IEEE Access 8. https://doi.org/10.1109/ACCESS.2020.2998358
25. Gartner identifies the top 10 strategic technology trends for 2023 (n.d.). Retrieved April 2, 2024, from https://www.gartner.com/en/newsroom/press-releases/2022-10-17-gartner-identifies-the-top-10-strategic-technology-trends-for-2023
26. Ghaffarianhoseini A, Tookey J, Ghaffarianhoseini A, Naismith N, Azhar S, Efimova O, Raahemifar K (2017) Building information modelling (BIM) uptake: clear benefits, understanding its implementation, risks and challenges. J Renew Sustain Energy Rev 75. https://doi.org/10.1016/j.rser.2016.11.083
27. Grieves M, Vickers J (2016) Digital twin: mitigating unpredictable, undesirable emergent behavior in complex systems. Transdiscipl Perspect Complex Syst New Find Approach. https://doi.org/10.1007/978-3-319-38756-7_4
28. Haag S, Anderl R (2018) Digital twin—proof of concept. J Manuf Lett 15. https://doi.org/10.1016/j.mfglet.2018.02.006
29. Hakimi O, Liu H, Abudayyeh O (2024) Digital twin-enabled smart facility management: a bibliometric review. Front Eng Manag 11(1). https://doi.org/10.1007/s42524-023-0254-4
30. Han B, Sun D, Yu X, Song W, Ding L (2020) Classification of urban street networks based on tree-like network features. Sustainability (Switzerland)12(2). https://doi.org/10.3390/su12020628
31. Javed AR, Shahzad F, Rehman S, ur, Zikria Y, Bin Razzak I, Jalil Z, Xu G (2022) Future smart cities requirements, emerging technologies, applications, challenges, and future aspects. Cities 129. https://doi.org/10.1016/j.cities.2022.103794
32. Jones D, Snider C, Nassehi A, Yon J, Hicks B (2020) Characterising the Digital Twin: a systematic literature review. CIRP J Manuf Sci Technol 29. https://doi.org/10.1016/j.cirpj.2020.02.002
33. Karimi H, Adibhesami MA, Hoseinzadeh S, Salehi A, Groppi D, Garcia DA (2024) Harnessing deep learning and reinforcement learning synergy as a form of strategic energy optimization in architectural design: a case study in Famagusta, North Cyprus. Buildings 14(5):1342. https://doi.org/10.3390/BUILDINGS14051342
34. Kavanagh JS (1989) Public streets for public use. Landsc J 8(1). https://doi.org/10.3368/lj.8.1.53
35. Kor M, Yitmen I, Alizadehsalehi S (2023) An investigation for integration of deep learning and digital twins towards construction 4.0. Smart Sustain Built Environ 12(3). https://doi.org/10.1108/SASBE-08-2021-0148
36. Kowalska-Pyzalska A, Byrka K (2019) Determinants of the willingness to energy monitoring by residential consumers: a case study in the city of wroclaw in poland. Energies 12(5). https://doi.org/10.3390/en12050907
37. Kritzinger W, Karner M, Traar G, Henjes J, Sihn W (2018) Digital twin in manufacturing: a categorical literature review and classification. IFAC-PapersOnLine 51(11). https://doi.org/10.1016/j.ifacol.2018.08.474
38. Leng J, Zhang H, Yan D, Liu Q, Chen X, Zhang D (2019) Digital twin-driven manufacturing cyber-physical system for parallel controlling of smart workshop. J Ambient Intell Humlzed Comput 10(3). https://doi.org/10.1007/s12652-018-0881-5

39. Liu M, Fang S, Dong H, Xu C (2021) Review of digital twin about concepts, technologies, and industrial applications. J Manuf Syst 58. https://doi.org/10.1016/j.jmsy.2020.06.017
40. Lock O, Pettit C (2022) Developing participatory analytics techniques to inform the prioritisation of cycling infrastructure. ISPRS Int J Geo Inf 11(2). https://doi.org/10.3390/ijgi11020078
41. Lofland LH (2017) The public realm: exploring the city's quintessential social territory. In: The public realm exploring the city's quintessential social territory. 10.4324/9781315134352
42. Lu Y, Liu C, Wang KIK, Huang H, Xu X (2020) Digital twin-driven smart manufacturing: connotation, reference model, applications and research issues. J Robot Comput Integr Manuf 61. https://doi.org/10.1016/j.rcim.2019.101837
43. McLaren Loring J (2007) Wind energy planning in England, Wales and denmark: factors influencing project success. Energy Policy35(4). https://doi.org/10.1016/j.enpol.2006.10.008
44. Mirror worlds, or, The day software puts the universe in a shoebox. . .: how it will happen and what it will mean (1992) Choice Rev Online 29(09). https://doi.org/10.5860/choice.29-5165
45. Morgan G, Yanckello R, Thayer T-L, Sheehan T, Farrell G, Mahmood S, Winckless C, MacDonald N (2022) Top technology trends in higher education for 2022. Technical report, Gartner
46. Nica E, Popescu G H, Poliak M, Kliestik T, Sabie O M (2023) Digital twin simulation tools, spatial cognition algorithms, and multi-sensor fusion technology in sustainable urban governance networks. Mathematics 11(9). https://doi.org/10.3390/math11091981
47. Ożadowicz A, Grela J (2017) Energy saving in the street lighting control system—a new approach based on the EN-15232 standard. Energy Effi10(3). https://doi.org/10.1007/s12053-016-9476-1
48. Qi Q, Tao F (2018) Digital twin and big data towards smart manufacturing and industry 4.0: 360 degree comparison. IEEE Access 6. https://doi.org/10.1109/ACCESS.2018.2793265
49. Qi Q, Tao F, Zuo Y, Zhao D (2018) Digital twin service towards smart manufacturing. Procedia CIRP 72. https://doi.org/10.1016/j.procir.2018.03.103
50. Santos C, Jimenez JA Espinosa F (2019) Effect of event-based sensing on IoT node power efficiency. Case study: air quality monitoring in smart cities. IEEE Access 7 https://doi.org/10.1109/ACCESS.2019.2941371
51. Sharma M, Tomar A, Hazra A (2024) Edge computing for industry 5.0: fundamental, applications and research challenges. IEEE IoT J
52. Tao F, Cheng J, Qi Q, Zhang M, Zhang H, Sui F (2018) Digital twin-driven product design, manufacturing and service with big data. Int J Adv Manuf Technol 94(9–12). https://doi.org/10.1007/s00170-017-0233-1
53. Tao F, Qi Q, Liu A, Kusiak A (2018) Data-driven smart manufacturing. J Manuf Syst 48. https://doi.org/10.1016/j.jmsy.2018.01.006
54. Tao F, Sui F, Liu A, Qi Q, Zhang M, Song B, Guo Z, Lu SCY, Nee AYC (2019) Digital twin-driven product design framework. Int J Prod Res 57(12). https://doi.org/10.1080/00207543.2018.1443229
55. Tao F, Zhang H, Liu A, Nee AYC (2019) Digital twin in industry: state-of-the-art. IEEE Trans Ind Inf 15(4). https://doi.org/10.1109/TII.2018.2873186
56. Vachalek J, Bartalsky L, Rovny O, Sismisova D, Morhac M, Loksik M (2017) The digital twin of an industrial production line within the industry 4.0 concept. In: Proceedings of the 2017 21st international conference on process control, PC 2017. https://doi.org/10.1109/PC.2017.7976223
57. Vickers R (2006) The digital studio—a holistic and student enabling learning environment. EISTA 2006 In: Malpica F, Tremante A, Welsch F (eds) Organizational EISTA '06: 4th international conference on education and information systems: technologies and applicat/soic'06: 2nd international conference on social and informatics and cybernetics, vol I, pp 33–35
58. Weil C, Bibri SE, Longchamp R, Golay F, Alahi A (2023) Urban digital twin challenges: a systematic review and perspectives for sustainable smart cities. Sustain Cities Soc 99. https://doi.org/10.1016/j.scs.2023.104862

59. Yu W, Patros P, Young B, Klinac E, Walmsley T G (2022) Energy digital twin technology for industrial energy management: classification, challenges and future. Renew Sustain Energy Rev 161. https://doi.org/10.1016/j.rser.2022.112407
60. Zhao Y, Wei Y, Zhang S, Guo Y, Sun H (2024) Multi-objective robust optimization of integrated energy system with hydrogen energy storage. Energies 17(5):1132. https://doi.org/10.3390/EN17051132

Digital Twin Applications

Applications of Digital Twin Technology in Consumer Energy Services

Ammar Al-Bazi and Mohammed Al-Khafajiy

Abstract This chapter intends to provide an overview of the role of the Digital Twin in Urban Planning. It explores the transformative potential of Digital Twin Technologies in addressing the multifaceted challenges of urban energy management and sustainability. It examines the requirements and implications of Digital Twin Technologies in consumer energy services. A hypothetical framework is proposed for monitoring and controlling energy services, integrating renewable energy sources with advanced technologies to optimise energy usage and promote sustainability in residential and urban areas.

Keywords Digital twin technologies · Consumer energy services · Urban planning · Energy management

1 Introduction

Governments and organisations worldwide are keen on addressing the multifaceted challenges arising from urbanisation. Issues such as urban poverty, escalating urban costs, electricity shortages and rising costs, traffic congestion, housing shortages, and environmental degradation is being exacerbated by urbanisation [1]. One prominent issue is consumer energy services and electricity/power generation/consumption estimation in urban areas [2]. In response to these challenges, the concept of the digital twin emerges as a potent tool to comprehensively model, simulate, and optimize urban environments. Thus, leveraging digital twins can aid in mitigating the adverse impacts of urbanisation and promoting sustainable urban growth.

Ammar Al-Bazi and Mohammed Al-Khafajiy, the authors, contributed equally to this work.

A. Al-Bazi (✉)
Aston Business School, Aston University, Birmingham, UK
e-mail: a.al-bazi@aston.ac.uk

M. Al-Khafajiy
School of Computer Science, University of Lincoln, Lincoln, UK
e-mail: malkhafajiy@lincoln.ac.uk

S. Pourroostaei Ardakani and A. Cheshmehzangi (eds.), *Digital Twin Computing for Urban Intelligence*, Urban Sustainability,
https://doi.org/10.1007/978-981-97-8483-7_8

Digital Twin Technology (DTT) is leading the way in the fourth and fifth industrial revolutions, representing a convergence of the physical and virtual worlds. A digital twin is a sophisticated simulation model that replicates a real-world object or system in a digital form, encompassing its properties, behaviours, and dynamics [3]. The origin of this innovation can be linked to NASA's pioneering use of pairing technology in intricate spacecraft simulations [4]. Today, it has evolved into a sophisticated tool that can perform real-time analysis and preemptive maintenance in various industries [5].

The structure of DTT consists of multiple essential parts that play a role in its operation. At its essence, it combines data from numerous sensors and sources to create a dynamic, real-time model of the physical object [6]. This information, collected via the Internet of Things (IoT), is examined using artificial intelligence (AI) to inform decisions and predict outcomes. Moreover, DTT can expand beyond traditional simulation boundaries by integrating advanced technologies like machine learning and big data analytics [7, 8]. Figure 1 shows an example of DT components consisting of three DDT environments processed through advanced models.

Although the energy industry receives significant advantages from DTT, its uses are widespread across various sectors. In manufacturing, digital twins optimise production lines and preempt equipment failures, whereas, in healthcare, they facilitate personalised medicine and advanced patient care [9, 10]. Urban planning also benefits cities by creating digital twins to streamline infrastructure development and enhance public services. These cross-industry applications underscore DTT's versatility and transformative potential, setting the stage for its pivotal role in the digital economy [11].

Fig. 1 DDT components and processes

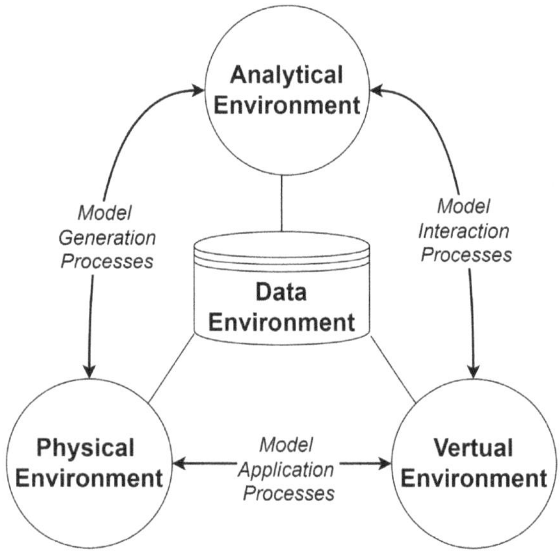

This chapter aims to contribute to the field of energy management for urban sustainability. It explores Digital Twin Technologies (DDT), understanding the foundational concepts and methodologies. It also sheds light on DDT requirements and the predictive modelling approaches of energy consumption. The practical implications of digital twin technology within consumer energy services are also discussed. It examines how digital twins can revolutionise energy distribution and load balancing, hence enhancing efficiency and reliability in energy delivery systems. Furthermore, a conceptual/hypothetical framework for monitoring and controlling energy services is presented to show the utilisation of the digital twin's potential for addressing the challenges posed by urbanisation and escalating energy demands. This proposes the integration of renewable energy sources with cutting-edge AI and digital twin technologies to optimise energy usage and promote sustainability in residential and urban settings. This integration aims to optimise energy usage patterns and ultimately promote long-term sustainability and resilience in energy management practices.

The rest of this chapter is structured as follows: an overview of Digital Twin Technologies, their requirements, and predictive modelling of energy consumption is presented in Sect. 2. The implications of digital twin technology in consumer energy services, along with the role of digital twins in energy distribution and load balancing are presented in Sect. 3. Section 4 presents a hypothetical framework for monitoring and control energy services using an advanced model for integrating renewable energy sources with cutting-edge AI and digital twin technologies to optimise energy use and sustainability in residential/urban areas. Finally, the paper is concluded in Sect. 5.

2 Digital Twin Technology Transformation

Digital Twin Technologies (DTT) have significantly revolutionised traditional approaches to energy management, contributing to urban intelligence and the cognitive intelligence of cities. The digital twins, sometimes called digital mirrors, seek to recreate intricate physical entities in a virtual digital environment by establishing a connection with real-world data [12]. This connection allows us to portray and simulate the real-time behaviour and characteristics of a physical system. This helps conduct a wide range of analysis and research within a virtual environment, which may be challenging or impractical to carry out directly in the real world. Additionally, digital twins support various advanced applications by providing a platform for experimentation and exploration. By creating virtual replicas of physical assets, processes, or systems within the energy sector, DTT enables real-time monitoring, analysis and optimisation, thereby enhancing the efficiency, sustainability, and reliability of energy initiatives, systems and technologies. This section explores the specific requirements, capabilities, and implications of DTT in energy management in greater detail.

2.1 Digital Twins Requirements for Energy Management

The effective utilisation of DTT in energy management necessitates the seamless integration of diverse data sources. These may include data from smart meters installed in residential and commercial buildings, sensors embedded within industrial equipment, controllers regulating energy flows and various Internet of Things (IoT) devices deployed across the energy infrastructure. Integrating these data streams provides a comprehensive view of energy consumption patterns, production levels, equipment performance, and environmental conditions. Through meticulous data integration, DTT constructs a digital representation of the physical energy ecosystem, enabling stakeholders to gain actionable insights and make informed decisions.

Focusing on DDT in smart energy engineering, it uses advanced physical models, data from smart sensors, operational plans, and maintenance logs to simulate various aspects of energy systems, like electricity, magnetism, heat, and fluid dynamics [12]. This simulation allows us to create a virtual reflection of smart energy systems in space. For example, a digital twin can mirror the entire lifespan of a smart device, constantly evolving and updating as needed. This ensures an accurate representation of the real energy system [13]. Moreover, digital twins enable precise interaction between *engineers/humans* and *computers*. Instructions from *users/experts* can directly affect physical equipment, and adjustments can be made to improve the simulation model over time. These interactions happen across different layers, as per Fig. 2, each serving its purpose in enhancing the simulation's accuracy/effectiveness and can be described as follows:

1. Physical or machinery layer: it represents the actual physical equipment/machines and systems residing in a real-world environment. It includes data sources such as sensors, actuators, and machinery involved in energy production, distribution, and consumption. The monitoring task for the status of conventional energy systems typically begin with the installation of sensors on energy equipment. However, gathering data from these sensors using data acquisition software can be challenging, especially when dealing with distributed/scattered data acquisition systems [14]. With the emergence of the Industrial Internet of Things (IIoT), a more advanced integration approach has been introduced and widely adopted to enrich data acquisition in the physical layer. IIoT technologies help with configuring a network of advanced and resilient sensors to collect a diverse range of data in different forms (types) from various devices within the energy system. This includes physical perception data (from ambient sensors), model generation data, and virtual-real fusion data. Integrating such massive data sources will enable the interconnecting of subsystems within the energy system, thereby enhancing monitoring and management capabilities.
2. Mechanism layer: this layer is located on the top of the physical layer. The mechanism layer interprets and implements instructions from the digital twin simulation. It controls how the physical equipment operates based on feedback from the simulation. It serves as the bridge between the physical equipment and the digital twin simulation. At this level, sophisticated algorithms and control mechanisms

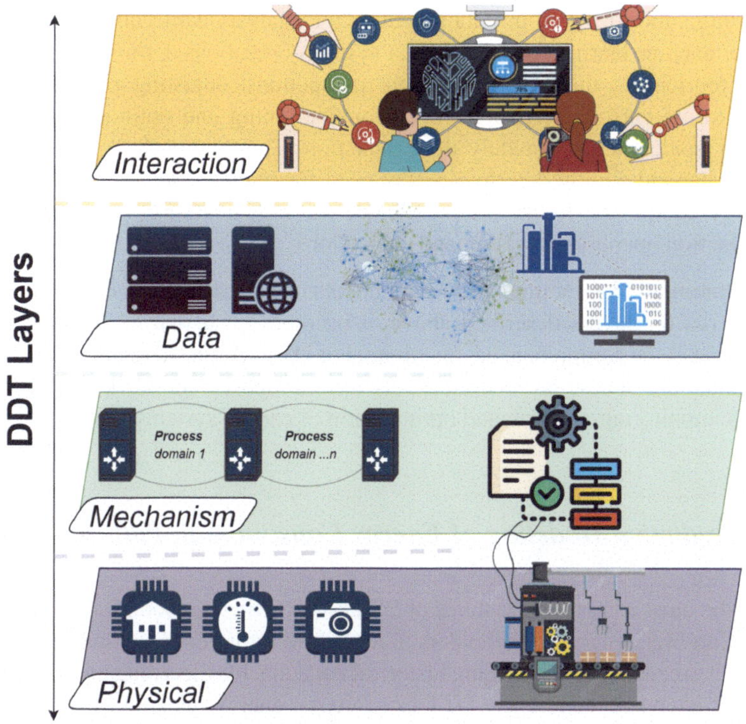

Fig. 2 DDT interactions happening across different layers

are employed to govern the operation of physical equipment based on feedback received from the simulation.

3. Data layer: the traditional energy system state monitoring primarily relies on basic sensory data and maintenance logs. In contrast, adopting DDT allows multi-dimensional data encompassing the entire life cycle of equipment and systems operations. In the data layer, the collected raw data from all devices in real-time, is normalised to ensure consistency and compatibility across different devices/systems. Data transmission within the data layer utilises low-delay communication channels to facilitate rapid and efficient transfer of information. Generally, Cloud/Edge computing is leveraged within the data layer to fulfil diverse computing and operational requirements. These technologies provide the necessary infrastructure for processing and analysing vast amounts of data generated by both digital twins and physical systems.
4. Interaction layer: at this level, there is a direct interference between humans/experts and the digital twin simulation, utilising the data at the data layer. Generally, a digital twins simulation model for intelligent energy systems constructed through a digital twin employs a hybrid modelling approach, which combines *model*-driven and *data*-driven methodologies. This method involves

the integration of data in data-pipelines for analysis while also incorporating deep/machine learning approaches for systems/operations optimisations. Human interventions at this layer, providing instructions, adjusting parameters, and monitoring performance, are crucial for fine-tuning and optimising the simulation models and ultimately the production. This approach holds significant value beyond simply determining the need for maintenance or replacement of energy equipment, it also offers valuable recommendations/predictions for the production operations and demand estimations.

Illustrating the interactions and requirements of digital twins in energy management across the layers described above, it becomes clear that each level plays a crucial role in the comprehensive functioning of DTT within smart energy engineering applications. These layers collectively form a robust framework that facilitates effective monitoring, control, and optimisation of energy systems.

2.2 Predictive Modeling of Energy Consumption Patterns

One of the most compelling features of DTT is its predictive modelling capabilities, which empowers energy stakeholders to anticipate future energy consumption patterns with precision. By leveraging historical data, machine learning algorithms, and advanced analytics techniques, DTT can forecast energy demand trends based on factors such as weather patterns, time of day, seasonal variations, and industrial processes. These predictive insights enable proactive energy management strategies, including optimised resource allocation, predictive maintenance scheduling, and demand response planning. Such predictive modelling includes a recommendation system for energy applications in:

1. **Demand-side energy management**: many research has been conducted on studying recent invitations and technologies for reducing electricity demand while also recommending patterns, for example, switching from $peak$ to $off\text{-}peak$ and vice versa. Such approaches require precise data to monitor and control household energy consumption, while also recommendation systems that provide energy efficiency tips to encourage behavioural changes for energy saving [15, 16]. Moreover, the DSM is a major area of interest for smart grid [17], with demand-side Personalised Recommendations System (PRS) changing users' perceptions and interests in their energy-saving habits and appliance usage [18]. Also, there is a vast amount of research work carried out on Demand Response (DR) based recommender systems for the energy supply chain [18, 19].
2. **Consumer behaviour modelling/recommendation**: the importance of mining consumer behavioural patterns from power system data are increasing, driven by the role of behaviour analysis and recommendation systems in promoting energy efficiency. Consumer behaviour is often characterised as soft and multifaceted attribute, which poses challenges in objective measurement compared to energy consumption measurement [20]. It is important to note that not all consumer

behaviours translate into energy conservation. Therefore, it is crucial to identify pertinent behavioural patterns that contribute to energy savings [20]. These significant patterns should manifest in actionable steps leading to energy conservation and encompass routine occurrences. The primary objective in this context is for recommender systems to identify instances where consumers deviate from established/noun patterns and suggest actions conducive to energy reduction. Traditionally, acquiring insights into human behaviour involves employing approaches such as price response replication, discrete choice models for technology adoption, and constant elasticity of substitution [21]. As indicated in [22], two-third of energy consumption in residents is associated with human behaviour. By identifying and understanding energy consumption patterns, these systems can offer tailored recommendations that not only optimise energy usage but also promote sustainable practices and reduce environmental impacts [23].

3 Digital Twins in Energy Demand-Side Management

3.1 Implications of Digital Twin Technology in Consumer Energy Services

The rise of DTT heralds a transformative shift in the energy sector, particularly in how energy providers engage with consumers and manage resources. By creating a dynamic digital representation of energy systems, DTT offers unparalleled insights into energy consumption patterns at an individual level [24]. These insights enable utility providers to personalise energy solutions, tailoring them to the unique demands of each consumer, which can lead to more efficient energy use and cost savings.

From the consumer perspective, DTT promotes an in-depth engagement with energy consumption. By visualising their consumption in real-time through digital twin interfaces, consumers can make informed decisions about their energy habits, potentially reducing their environmental impact, contributing to broader energy conservation efforts and understanding their daily habits [25, 26]. On the other hand, energy producers utilise DTT technologies to monitor customers' appliances for quality compliance, customer service, and diagnosis reports. See Fig. 3.

Moreover, integrating DTT with smart home technology marks a significant advance. Digital twins can orchestrate a symphony of smart devices, optimising their operation to reduce unnecessary power usage without compromising comfort or convenience. This level of automation and predictive maintenance enhances energy efficiency and extends the lifespan of devices by anticipating and addressing potential failures before they occur [27]. In a broader sense, the implications of DTT extend to the stabilisation and optimisation of the power grid itself. By simulating various consumption scenarios, energy providers can better plan for and react to fluctuating demand [26]. This includes the continuous integration of renewable energy sources

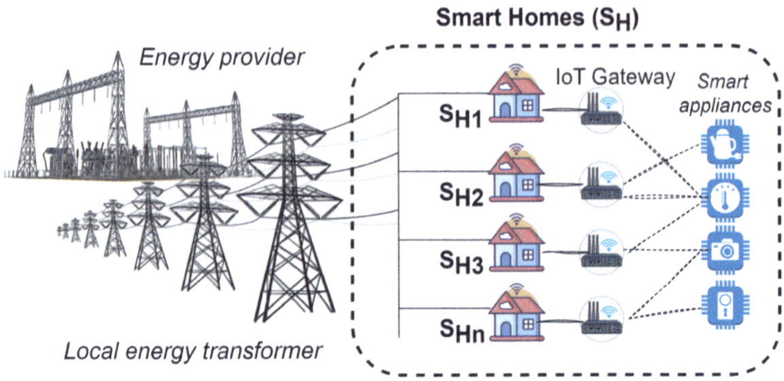

Fig. 3 DTT model linking electric appliances with energy producers through IoT

known for their variable output. DTT can help manage these variations, ensuring a steady power supply and facilitating the transition to greener energy sources.

3.2 The Importance of Digital Twin Technology in Consumer Energy Services

Digital Twin Technology (DTT) is quickly becoming a game-changer in consumer energy services, ushering in a new age of personalised and efficient energy management. The application of DTT enables the creation of personalised energy solutions that transform consumer experiences. Utilising real-time data, digital twins predict and adapt to individual energy usage patterns, facilitating the design of bespoke energy optimisation strategies for consumers [28]. This personalisation enhances energy efficiency and paves the way for consumers to play a more proactive role in managing their energy consumption [29].

Central to the value proposition of DTT is its capacity to promote increased consumer engagement. Digital twins serve as interactive platforms that provide users with deep insights into their energy usage through intuitive visualisations and simulations [6]. This increased level of transparency gives consumers the knowledge to make informed decisions, promoting energy conservation and drive cost savings [30]. Furthermore, DTT allows energy service providers to offer dynamic pricing models and personalised energy plans. It encourages consumers to adjust their consumption to less busy periods, ultimately decreasing costs and supporting grid reliability [26].

The combination of DT technology and smart home technology marks a significant advancement in consumer energy services. Digital twins work harmoniously with smart home systems, coordinating multiple IoT devices to provide automated energy-saving tasks and anticipate maintenance needs [31]. This integration ensures that energy efficiency is not just a manual task but a continuous, automated process, optimising energy consumption without compromising comfort or convenience.

From smart thermostats to intelligent lighting systems, DTT is the invisible hand guiding the smart home towards sustainable living [32].

3.3 The Role of Digital Twins in Energy Distribution and Load Balancing

DDT is driving a major transformation in the field of energy distribution and load management. DTT optimises grid operations by incorporating predictive analytics and real-time monitoring [33]. Digital twins act as a neural network across the energy grid, continuously analysing data to predict demand patterns, foresee potential disruptions, and recommend optimal energy distribution [34]. This proactive approach to grid management not only improves reliability but also maximises efficiency of energy resources.

In the context of renewable energy integration, the variable and intermittent nature of sources like solar and wind pose unique challenges. DTT is an invaluable tool in this landscape by simulating various renewable energy outputs under different conditions. These simulations enable energy providers to manage the variability and ensure a steady energy supply by integrating renewables into the grid more effectively [35]. This results in a more resilient grid system that can adapt to the fluctuating nature of renewable energy sources while meeting the demand with a lower carbon footprint [36].

DTT also transforms demand response strategies by allowing for more sophisticated and efficient real-time adjustments to energy supply. Digital twins offer a detailed view of consumption trends and allow the grid to respond dynamically to changes in demand. This responsiveness is crucial during peak load periods, emergencies, or when integrating demand-side resources [37]. By simulating and adjusting to these scenarios in real-time, digital twins aid utilities in smoothing out demand curves, reducing strain on the grid and ultimately driving down costs for both providers and consumers [6]. Overall, the role of Digital Twin Technology in modernising energy distribution systems is complex and significant. It makes energy grids are smarter and more responsive and support the seamless incorporation of renewable energies, paving the way for a more sustainable and efficient energy landscape.

3.4 Efficient Energy Utilisation (Storage Batteries Allocation and Location)

Implementing Digital Twin Technology (DTT) in the energy sector has brought about a paradigm shift in the analysis and optimisation of energy consumption. DTT provides a detailed perspective of individual consumer energy patterns, leading to significant enhancements in identifying and implementing efficiency improvements.

By mirroring the energy consumption behaviours of consumers, digital twins enable utility providers to pinpoint inefficiencies and anomalies [26]. This in-depth analysis facilitates personalised energy-saving recommendations, paving the way for more sustainable consumption practices.

Beyond consumption patterns, DTT is instrumental in augmenting the performance of energy storage systems, such as batteries. These digital replicas provide a strategic platform for managing storage unit charge and discharge cycles, ensuring optimal usage. By utilising predictive algorithms, DTT can effectively smooth out peaks in energy demand, thereby reducing the reliance on less sustainable energy sources during high-demand periods [38]. This capability is crucial for enhancing the integration of renewable energy sources into the grid, as it offsets their natural fluctuations and promotes a more resilient energy infrastructure.

DTT also excels in the simulation of complex demand scenarios. Creating virtual replicas of energy systems allow stakeholders to test and evaluate the impact of various demand patterns and extreme conditions. This insight is vital for disaster response, where rapid, informed decision-making can mitigate risks and maintain the energy supply [39]. Moreover, capacity planning benefits from these simulations by providing a platform for the future scaling of energy resources, ensuring that the infrastructure evolves in sync with demand projections.

DTT is a fundamental building block in progressing toward smarter and more adaptable energy systems. By providing deep insights into consumption, optimising energy storage, and enabling robust scenario planning, Digital Twin Technology ensures that energy utilisation is more efficient and ready to meet the demands of tomorrow's challenges [40].

4 Hypothetical Framework for Monitoring and Control Energy Services

This hypothetical framework is designed as a sophisticated approach to energy management. It uses technology and data to make informed decisions that benefit both energy suppliers and consumers, ensuring dynamic and efficient energy distribution. This framework is an advanced model for integrating renewable energy sources with cutting-edge AI and digital twin technologies to optimise energy use and sustainability in residential areas. See Fig. 4 for the hypothetical framework.

This framework integrates energy suppliers, a virtual simulation environment (Digital Twin Technology), real-world home energy management (HEM), and weather data, all coordinated through AI (Artificial Intelligence) technologies. The detailed breakdown of each component is provided as follows:

1. Energy Supplier This module shows multiple energy suppliers (S1, S2, ... Sn), each with different sources such as solar and wind turbine resources. This suggests that the energy supplied is renewable, likely solar and/or wind energy. These suppliers are feeding energy into real-world household systems.

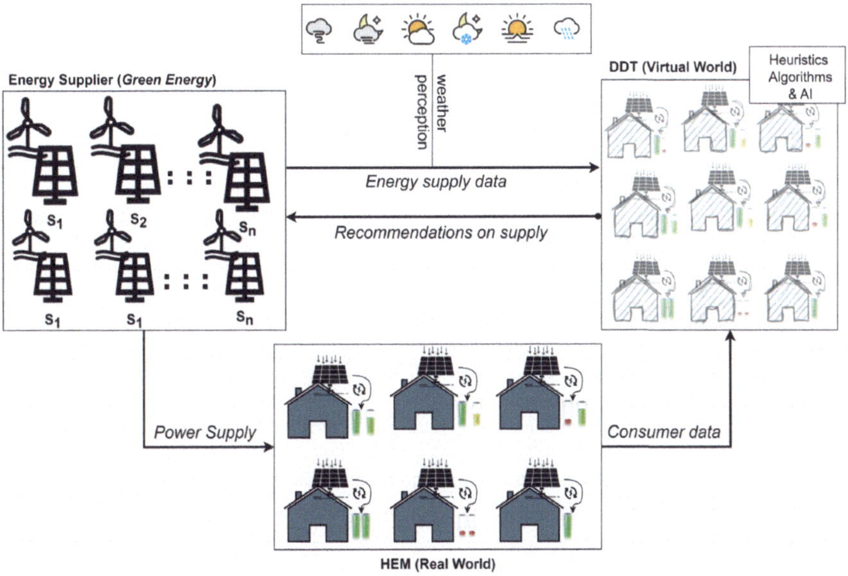

Fig. 4 Hypothetical Household Intelligent Energy Management System

2. DDT (Virtual World) The Digital Twin (DDT) module presents a virtual simulation of a residential area, where each house is equipped with solar panels and different storage devices. This virtual model simulates and analyses energy consumption and production scenarios. Each house storage device displays different charging statuses indicated by green or red lights, possibly representing energy surplus or deficit.
3. Weather Data This part of the framework provides weather conditions that likely influence energy production and consumption, such as sunny, cloudy, and rainy conditions. This data is crucial for the virtual and real-world models to simulate or adjust energy management strategies accurately.
4. HEM (Real World) The Home Energy Management (HEM) module represents actual houses in the real world. These houses also have solar panels and energy storage devices, and similar to the virtual model, they show different energy statuses with green and red indicators. This suggests active management of energy based on consumption needs and production capacity.
5. Heuristics Algorithm and AI Integration: At the centre of the framework, an AI symbol indicates that artificial intelligence algorithms are used to process data from the weather inputs to optimise energy distribution and consumption, improve efficiency, and possibly predict future energy needs or production rates based on weather forecasts and other variables. The heuristics algorithm predicts optimal energy distribution strategies and simulates future supply scenarios, facilitating informed decisions with suppliers to maximise efficiency and meet household demand.

Algorithm 1 Urban Energy Management Framework

1: At the core, multiple energy suppliers ($S_1, S_2, ..., S_n$) involve renewable sources like solar (P_{Solar}) and wind turbines (P_{Wind}) alongside traditional power plants ($P_{\text{Traditional}}$).
2: Each house or building in a district is equipped with Roof-Mounted Solar Panels and storage devices (S) of varying sizes.
3: Houses report their usage (U) and storage device status (S_{status}) to a cloud, which sends real-time data to a virtual simulation module.
4: The virtual simulation module utilizes DDT to mirror houses and predict optimal energy distribution strategies using heuristic algorithms.
5: DDT adjusts energy production and distribution forecasts based on real-time data and weather profiles using AI.
6: Predication can be $E_{\text{forecast}} = f(U, S_{\text{status}}, P_{\text{Solar}}, P_{\text{Wind}}, P_{\text{Traditional}}, \text{Weather})$.
7: AI analyzes energy consumption patterns (C), supplier output (P_{Supplier}), and weather forecasts (W) to enhance decision-making processes and improve future predictions, where $E_{\text{optimal}} = g(C, P_{\text{Supplier}}, W)$.
8: Integration of AI within DDT enhances adaptability, efficiency, and sustainability in urban energy management.
9: AI-powered analytics optimize energy output, minimize wastage, and improve overall system performance, contributing to resilient and sustainable urban energy systems.

Algorithm 1 highlights the main logical steps of the proposed framework and can be summarised as follows: At the framework core, multiple energy suppliers (S1, S2, Sn) involve renewable sources like solar and wind turbines, alongside other traditional power plants, which secure the required energy to power a network of homes or other buildings. Each house or building in a district is equipped with a Roof-Mounted Solar Panels and a number of storage devices. The sizes of these houses, buildings, storage devices and energy consumption differ. The energy deficit could occur due to many reasons, including but not limited to weather profile, excess use of demand, and the used storage device capacity. These houses report their usage and storage devices' status to a cloud that, in turn, sends the data on a real-time basis to a virtual simulation module.

This module includes DDT, a virtual simulation environment that mirrors the houses, including their solar energy and storage batteries. Utilising heuristics algorithms, the DDT model predicts optimal energy distribution strategies based on real-time data and simulated future scenarios, facilitating informed decisions by exchanging information with suppliers to communicate with houses, maximise their efficiency, and meet their demand by providing the deficit energy with less energy supplied from suppliers and minimised costs.

Moreover, Artificial Intelligence (AI) is employed with the DDT model to analyse data from weather profiles to adjust energy production and distribution forecasts, ensuring adaptability to upcoming weather conditions and optimising energy output. This includes energy consumption patterns, supplier output, weather forecasts, enhancing decision-making processes, and continuously learning from historical data to improve future predictions and operations. The integration of AI within the DTT model enhances adaptability, efficiency, and sustainability in urban energy management. Leveraging AI-powered analytics with DTT allows the system to optimise energy output, minimise wastage, and improve overall system performance, ultimately contributing to more resilient and sustainable urban energy systems.

5 Conclusion

This chapter highlights the transformative potential of Digital Twin Technologies in urban energy management. By enabling demand monitoring, predictive modelling of energy consumption and optimising energy distribution, DTT can offer robust solutions to challenges posed by urbanisation. Integrating renewable energy sources with DTT can further enhance sustainability. The future work for us is primarily to focus on the development and implementation of the hypothetical framework proposed in this chapter. This involves designing and implementing a robust system for monitoring and controlling energy services using an advanced model that integrates renewable energy sources with cutting-edge AI and DT technologies. This includes refining the algorithms and methodologies used for energy optimisation and integrating real-time data analytics capabilities to enable proactive decision-making. This framework can potentially contribute to more sustainable and efficient urban energy management, hence benefiting urban planning.

References

1. O'Brien P, Pike A, Tomaney J (2019) Governing the 'ungovernable'? financialisation and the governance of transport infrastructure in the London 'global city-region'. Progress Plan 132:100422
2. Ayala-Chauvin M, Riba Sanmartí G, Riba C, Lara P (2022) Evaluation of the energy autonomy of urban areas as an instrument to promote the energy transition. Energy Sources, Part B: Econ Plan Policy 17(1):2053897
3. Neethirajan S, Kemp B (2021) Digital twins in livestock farming. Animals 11(4):1008
4. Kaiblinger A, Woschank M (2022) State of the art and future directions of digital twins for production logistics: a systematic literature review. Appl Sci 12(2):669
5. Ke G, Dashun Z, Liqin M, Jing W, Juan D, Shubo L, Huan Q, Lizhong S (2023) Intelligent machine plus production line digital twin model construction technology. J Phys Conf Ser 2478:102011. IOP Publishing
6. Fuller A, Fan Z, Day C, Barlow C (2020) Digital twin: enabling technologies, challenges and open research. IEEE Access 8:108952–108971
7. Madni AM, Madni CC, Lucero SD (2019) Leveraging digital twin technology in model-based systems engineering. Systems 7(1):7
8. Li L, Mao C, Sun H, Yuan Y, Lei B (2020) Digital twin driven green performance evaluation methodology of intelligent manufacturing: hybrid model based on fuzzy rough-sets AHP, multistage weight synthesis, and PROMETHEE II. Complexity 2020:1–24
9. Bécue A, Maia E, Feeken L, Borchers P, Praça I (2020) A new concept of digital twin supporting optimization and resilience of factories of the future. Appl Sci 10(13):4482
10. Croatti A, Gabellini M, Montagna S, Ricci A (2020) On the integration of agents and digital twins in healthcare. J Med Syst 44(9):161

11. Rasheed A, San O, Kvamsdal T (2020) Digital twin: values, challenges and enablers from a modeling perspective. IEEE Access 8:21980–22012
12. Huang W, Zhang Y, Zeng W (2022) Development and application of digital twin technology for integrated regional energy systems in smart cities. Sustain Comput Inf Syst 36:100781
13. Misyurin SY, Kreinin GV, Nosova NY (2021) Digital twin of the drive system, considering the forces of various nature. Procedia Comput Sci 190:611–621; In: 2020 annual international conference on brain-inspired cognitive architectures for artificial intelligence: eleventh annual meeting of the BICA society. https://doi.org/10.1016/j.procs.2021.06.072
14. Wang XV, Wang L (2019) Digital twin-based WEEE recycling, recovery and remanufacturing in the background of industry 4.0. Int J Prod Res 57(12):3892–3902
15. Alsalemi A, Sardianos C, Bensaali F, Varlamis I, Amira A, Dimitrakopoulos G (2019) The role of micro-moments: a survey of habitual behavior change and recommender systems for energy saving. IEEE Syst J 13(3):3376–3387
16. Onile AE, Machlev R, Petlenkov E, Levron Y, Belikov J (2021) Uses of the digital twins concept for energy services, intelligent recommendation systems, and demand side management: a review. Energy Rep 7:997–1015
17. Metwaly MK, Teh J (2020) Probabilistic peak demand matching by battery energy storage alongside dynamic thermal ratings and demand response for enhanced network reliability. IEEE Access 8:181547–181559
18. Luo F, Ranzi G, Kong W, Dong ZY, Wang S, Zhao J (2017) Non-intrusive energy saving appliance recommender system for smart grid residential users. IET Gener Transm Distrib 11(7):1786–1793
19. Behl M, Smarra F, Mangharam R (2016) Dr-advisor: a data-driven demand response recommender system. Appl Energy 170:30–46
20. Zehnder M, Wache H, Witschel H-F, Zanatta D, Rodriguez M (2015) Energy saving in smart homes based on consumer behavior: a case study. In: 2015 IEEE first international smart cities conference (ISC2). IEEE, pp 1–6
21. Venturini G, Tattini J, Mulholland E, Gallachóir BÓ (2019) Improvements in the representation of behavior in integrated energy and transport models. Int J Sustain Transp 13(4):294–313
22. Kashani A, Ozturk Y (2017) Residential energy consumer behavior modification via gamification. In: 2017 IEEE 6th international conference on renewable energy research and applications (ICRERA). IEEE, pp 1221–1225
23. Zhang Y, Meng K, Kong W, Dong ZY (2018) Collaborative filtering-based electricity plan recommender system. IEEE Trans Ind Inf 15(3):1393–1404
24. Moyne J, Qamsane Y, Balta EC, Kovalenko I, Faris J, Barton K, Tilbury DM (2020) A requirements driven digital twin framework: specification and opportunities. IEEE Access 8:107781–107801
25. Schleich B, Anwer N, Mathieu L, Wartzack S (2017) Shaping the digital twin for design and production engineering. CIRP Annals 66(1):141–144
26. Mansour D-EA, Numair M, Zalhaf AS, Ramadan R, Darwish MM, Huang Q, Hussien MG, Abdel-Rahim O (2023) Applications of IoT and digital twin in electrical power systems: a comprehensive survey. IET Gener Transm Distrib 17(20):4457–4479
27. Zhang H, Ren D, Ming H, Zhang W, Cao G, Liu J, Wang L, Song J, Qiu J, Wang J et al (2023) Digital twin enables rational design of ultrahigh-power lithium-ion batteries. Adv Energy Mater 13(1):2202660
28. Zhang Z, Zou Y, Zhou T, Zhang X, Xu Z (2021) Energy consumption prediction of electric vehicles based on digital twin technology. World Electr Veh J 12(4):160
29. Steindl G, Stagl M, Kasper L, Kastner W, Hofmann R (2020) Generic digital twin architecture for industrial energy systems. Appl Sci 10(24):8903
30. Kasper L, Birkelbach F, Schwarzmayr P, Steindl G, Ramsauer D, Hofmann R (2022) Toward a practical digital twin platform tailored to the requirements of industrial energy systems. Appl Sci 12(14):6981
31. Leitao J, Gil P, Ribeiro B, Cardoso A (2020) A survey on home energy management. IEEE Access 8:5699–5722

32. Lin C (2024) An intelligent illumination system with personalized quality of services in smart home. In: Ninth international symposium on sensors, mechatronics, and automation system (ISSMAS 2023), vol 12981. SPIE, pp 195–202
33. El Marai O, Taleb T, Song J (2020) Roads infrastructure digital twin: a step toward smarter cities realization. IEEE Netw 35(2):136–143
34. Kukushkin K, Ryabov Y, Borovkov A (2022) Digital twins: a systematic literature review based on data analysis and topic modeling. Data 7(12):173
35. Yu P, Zhaoyu W, Yifen G, Nengling T, Jun W (2023) Application prospect and key technologies of digital twin technology in the integrated port energy system. Front Energy Res 10:1044978
36. Paget M, Seacrest T, Widergren S, Balducci P, Orrell A, Bloyd C (2011) Using smart grids to enhance use of energy-efficiency and renewable-energy technologies
37. Ivanov D, Dolgui A (2021) A digital supply chain twin for managing the disruption risks and resilience in the era of industry 4.0. Prod Plan Control 32(9):775–788
38. Huang H, Wang H, Cai Y, Chen X, Li T (2022) Optimal configuration of fire-storage capacity considering dynamic charge-discharge efficiency of hybrid energy storage. Front Energy Res 10:950521
39. Dembski F, Wössner U, Letzgus M, Ruddat M, Yamu C (2020) Urban digital twins for smart cities and citizens: The case study of Herrenberg, Germany. Sustainability 12(6):2307
40. Deng M, Menassa CC, Kamat VR (2021) From BIM to digital twins: a systematic review of the evolution of intelligent building representations in the AEC-FM industry. J Inf Technol Constr 26

Integrating Digital Twins in Urban Sustainability: A Framework for University Campus Applications

Frederico Fialho Teixeira, Islam Mashaly, Maryam Shafiei, Qianbin Xu, Guanqi Zhu, and Jurij Karlovsek

Abstract In the contemporary urban environment, the convergence of smart technologies, virtual innovations, and the Internet of Things (IoT) has catalysed the need for robust frameworks that enhance data collection, visualisation, and analysis. Digital twins, scalable from individual buildings to entire cities, integrate diverse data sources to offer critical insights into facility management, asset lifecycle, and the optimisation of design and performance. While the application of digital twins has been explored extensively over the past two decades, there remains a significant gap in comprehensive frameworks that facilitate their development and implementation across various scales. This chapter introduces a systematic framework tailored for the creation of a digital twin for university campuses, which, with adaptations, can be extended to broader urban contexts such as towns and cities. This framework leverages underutilised data, employing advanced interpolation and visualisation techniques to generate analysis-ready datasets for simulations (e.g., flooding, energy consumption) and spatial mappings of functionality and occupancy. Such integrative approaches not only enhance campus planning and policymaking but also serve as a model for urban sustainability efforts. Further, this research delineates the specific methodologies for data extraction, transformation, and loading that are critical to optimising the digital twin framework at a campus level. The outcomes presented herein aim to guide future large-scale digital twin projects and provide urban and architectural researchers with a novel perspective on the potential of digital twins in future design, management, and sustainable planning. This includes predictive analytics

F. F. Teixeira (✉) · Q. Xu · G. Zhu
School of Architecture, Design and Planning, The University of Queensland, Brisbane, Australia
e-mail: f.frederico@uq.edu.au

I. Mashaly
School of Sciences & Engineering, American University in Cairo, Cairo, Egypt

M. Shafiei
Department of Architecture, American University in Dubai, Art & Design, Dubai, UAE

J. Karlovsek
School of Civil Engineering, The University of Queensland, Brisbane, Australia

S. Pourroostaei Ardakani and A. Cheshmehzangi (eds.), *Digital Twin Computing for Urban Intelligence*, Urban Sustainability,
https://doi.org/10.1007/978-981-97-8483-7_9

for problem-solving, risk assessment, cost management, operation and maintenance, and performance enhancement in urban settings.

Keywords Digital twin · Campus planning · Data integration · Sustainable development · Spatial information systems · Internet of Things · Urban design management

1 Introduction

The concept of digital twins, as explored by Broo and Schooling [3], signifies a ground breaking shift towards more efficient infrastructure and building management. These technological constructs not only mirror physical assets digitally but also amalgamate extensive data sets, offering a dynamic and thorough asset perspective. This comprehensive view is instrumental in refining decision-making and optimising resource deployment, echoing Hakimi et al. [5] findings. Through such advanced replication and data integration, stakeholders are equipped with unparalleled insights, facilitating proactive maintenance, operational efficiency, and the promotion of sustainability within facility management practices.

For the development and application of digital twins across different scales, from comprehensive urban developments to precise architectural endeavours, it is crucial to emphasise the need for advanced frameworks. These frameworks, essential for the effective amalgamation, visualisation, and interpretation of data, pave the way for significant advancements in facility management, asset lifecycle extension, and design performance enhancements. Deren et al. [4], Shahzad et al. [13], and Vivi et al. [16] have all highlighted the critical role of these frameworks in harnessing data from various sources to unveil key insights for planning and management. Despite the evident enthusiasm and recognised potential surrounding digital twins, research predominantly targets specific sectors, leaving a noticeable void in universal, interdisciplinary frameworks capable of spanning across industries [5]. This observation underscores a critical need for expansive research and development efforts aimed at bridging this gap, fostering a more inclusive approach to digital twin technology application.

This paper proposes an innovative framework for integrating digital twins into university campus management, potentially extending its application to a variety of built environments. At its core, this framework is supported by a data maturity model designed to address the complexities inherent in digital twin development, tailored to meet the specific needs and capabilities of diverse stakeholders. This approach goes beyond technical implementation, leveraging the complex networks within university settings to explore new frontiers in technological innovation. University campuses, characterised by their diverse array of buildings, infrastructure, and stakeholders, offer an ideal environment for the practical application and refinement of this framework (see Fig. 1). Choosing university campus as the case study allows for the integration of vast data sets from numerous sources, setting the stage for a comprehensive

Fig. 1 Diversity of components and categories that might be found in a university campus (*Source* Image created by the authors)

exploration of digital twin technology's potential to transform campus management and operations.

The creation of a digital twin for university campuses is not solely an exercise in technological prowess but a venture into the intersection of technology, academia, and community engagement. This approach underscores the value of digital twins in streamlining campus operations, audit transparency, enhancing environmental sustainability, and promoting a culture of innovation and collaboration. The envisioned framework not only promises to optimise energy use, waste management, and spatial planning, but also aims to galvanise the campus community, fostering a shared commitment to sustainability and technological advancement.

The presence of data collecting and smart technologies, the development of virtual technologies, and the emergence of the internet of things (IoT) have prompted the development of frameworks and efficient workflows for data collection, visualisation, and analysis [13, 16]. Digital twins in different scales from the city to the building, allow for bringing together data from different sources to generate fundamental and illuminating insights for the management of current facilities and the lifecycle of amenities, as well as the improvement of the performance of current and future designs [4]. This allowed research interest in digital twins to grow over the past decade [5]. While these studies demonstrated the potential benefits of using digital twins in specific industries, they did not provide a comprehensive framework for the development and use of digital twins that can be demonstrated in other settings across their specific industry. Most of these studies have focused on specific applications or industries, rather than providing a comprehensive framework or workflow for the development of digital twins and its applications [5].

The chapter meticulously unfolds the multifaceted role of digital twins in urban and architectural domains. A thorough literature review sheds light on the existing body of work, spotlighting the diverse methodologies and challenges encountered. The methodology section delineates the comprehensive approach employed in developing and validating the proposed digital twin framework for university campuses, detailing the integration and visualisation techniques utilised. The results and discussion section delves into the outcomes of a university campus case study, showcasing the framework's core components, its versatility across different settings, and its potential for scalability. Conclusively, the chapter reflects on the significant implications of these findings for the advancement of digital twins in the built environment and charts a course for future endeavours in asset information modelling. Digital twins have the potential to revolutionise the way we manage, operate, and maintain buildings and infrastructure [3]. They assist in creating digital replicas of physical assets and integrating various data sources that can provide a comprehensive and dynamic view of assets, enabling more informed decision-making and optimisation of resource utilisation [5].

The chapter, therefore, proposes a framework for creating digital twins for university campuses that could be applied to other buildings, infrastructure, and large scale-built environments in the future [18]. The framework will be characterised by a data maturity model that will help guide the development of any digital twins, considering the needs and capabilities of stakeholders. The creation of a university campus digital twin framework is not just a mere technical exercise, but rather a vital pursuit that leverages the potential of these complex systems to push the boundaries of what's possible. With their myriad of diverse buildings, infrastructure and stakeholders, university campuses provide a fertile ground to trial and fine-tune a digital twin framework that can effectively process and assimilate massive amounts of data from multiple sources (see Fig. 1). By developing a framework that takes into consideration the unique needs of all stakeholders, researchers can increase the likelihood of widespread adoption and usage, unlocking the full potential of the digital twin as a tool for enhancing campus management and operations. Additionally, university campuses present a thriving hub of research and innovation, affording researchers the opportunity to explore novel applications of digital twins, which can ultimately help drive technological progress and improvements in the built environment.

The research starts by providing a comprehensive overview of the existing research on digital twins, with a focus on studies that have addressed the development and use of digital twins in the built environment such as university campuses and urban areas. Subsequently, the study includes a discussion of the different approaches that have been proposed, as well as the challenges in each previous case studies. Subsequently, the methodology section presents the approach taken in this study to develop and test the proposed framework for digital twins with a focus on university campuses.

The framework presents a description of all data sources and tools used, as well as the steps taken to integrate and visualise all the information. To conclude, the results and discussion section presents the findings of the UQ university campus case study. This includes the key components of the proposed framework and how they can be applied to other built environments and how can it be scaled up. Furthermore, the findings will be presented so that it can inform the future of digital twins in the built environment, as well as how the proposed framework can be adapted to fit other contexts.

Overall, the aims and objectives of the research focus on the development a comprehensive framework for creating and implementing digital twins specifically tailored for university campuses. This framework utilises advanced data interpolation and visualization techniques to generate datasets for simulations, to enhance campus planning, policymaking, and overall sustainability efforts. The research aims to fill the existing gap in scalable digital twin frameworks and provide methodologies for data management that can be adapted to broader urban contexts. The main objective is to guide future large-scale digital twin projects, offering predictive analytics for problem-solving, risk assessment, cost management, and performance enhancement in urban settings.

2 Literature Review

The development of digital twins for the built environment, including university campuses and cities, presents several unique challenges and opportunities. To effectively address these challenges and opportunities, it is important to carefully consider a range of topics, including data management and integration, architecture and interoperability, legal and ethical considerations, and stakeholder engagement and participation.

The literature review analyses approaches and methods used in these studies, to detect opportunities for creating a generalised framework for digital twins to be applied to a variety of settings and industries. This becomes a foundation for the development of a digital twin framework for university campuses serving as a testbed for fine tuning and improving the framework for use in other contexts.

In the study conducted by Oliveira [10], the intent was to transform an airport using digital technologies. The industry's focus was on airports, and the scale of the study was comparable to that of a small city. The data used included information on working processes, approaches, and take-offs, as well as and building systems. This study involved prediction and monitoring of the data. In contrast, Alonso et al. [1] conducted research that focused on automating the construction of Building Information Modelling (BIM) digital objects using manufacturing processes. The data included quantities, cost, and workflow information. Tools utilised comprised Dynamo, SQL, and spreadsheets. This study did not involve prediction, simulation, or optimisation. However, it did include monitoring of quantities and optimising the manufacturing workflow based on the monitoring process. The study conducted by Alonso et al. [1] aimed to improve energy performance from construction to operation using a Platform as a Service (PaaS) by including data on life cycle cost. The industry focus was on residential buildings at a building-level scale. Building Information Modelling (BIM) and blockchain technology were utilised to monitor and optimise the energy performance.

The study conducted by Zaballos et al. [18] delved into the level of comfort within a university campus building. The research was conducted at the building level, using environmental monitoring and emotion detection data. To gather this data, Zaballos et al. utilised Building Information Modelling (BIM) and Wise (IoT) technology. On the other hand, Wang et al. [17] aimed to assess efficiency of digital twin sensor technology by analysing data on environmental satisfaction. Their tools of choice included a Data Fusion Algorithm (DFA) and deep machine learning. This study did not involve prediction or optimisation, but it did include simulation and monitoring. However, Life Cycle (LC) data and sensing were not a primary focus of this study. Both studies utilised a variety of methods and techniques, including BIM, Dynamo, SQL, spreadsheets, data fusion algorithms, deep machine learning, and the Internet of Things (IoT).

Han et al. [6] propose the use of digital twin technology, guided by the theoretical model proposed by the digital twin technology, using UAV tilt photography and 3D modelling to collaboratively build the virtual campus scene. They accurately map the physical campus onto the virtual model with real-time sensing, and remotely control it to achieve the reverse control of the twin virtual campus to the physical campus. The modules of the smart campus system were evaluated using a System Usability Scale based on student experiences. The experimental results demonstrated that the virtual-real campus system can enhance school management and teaching, providing important implications for promoting the development and application of campus intelligent systems.

A review of previous research highlights a gap in the development of a comprehensive framework for digital twins that can be applied to various industries and scales. Some studies, such as those by Oliveira [10] and Alonso et al. [1], concentrate on specific industries or scales. However, there is a need to a more universally applicable framework within the built environment.

2.1 Standards in Place

Recent development of ISO 23247, Digital Twin Manufacturing Framework, provides a general development framework that may be instantiated for case-specific implementations of digital twins in manufacturing [14]. The standard consists of four parts: (1) General principles and overview, (2) Reference architecture, (3) Digital representation, and (4) Information interchange.

According to ISO 23247, the framework is particularly designed for Manufacturing processes and equipment. Nonetheless, ISO 23247 holds potential advantages for the development of a digital twin framework for the built environment as well.

Firstly, it offers a comprehensive definition of what constitutes a digital twin, and how it can be applied, which can assist in establishing a universal understanding of the concept between researchers and practitioners. Moreover, the framework establishes key components that are necessary for the creation and utilisation of digital twins in manufacturing, such as data management, interoperability, security, and data analytics. These components can also be incorporated into the development of digital twins for the built environment, ensuring that the digital twin is trustworthy, secure, and capable of providing valuable insights. Furthermore, the framework provides guidelines for the management and preservation of digital twins over time, proving to be helpful in maintaining the digital twin's accuracy and up-to-date status as the physical object or system changes.

Although the ISO 23247 framework can be an essential resource for those seeking to develop a digital twin framework for the built environment, there are some limitations to consider. It is based on a centralised model of data management and control, which might not be suitable for the built environment, which is characterised by a distributed and decentralised nature Broo et al. [3]. This may necessitate further considerations to make sure that the digital twin can effectively incorporate data from various sources and stakeholders. Additionally, the framework might not fully address the legal and ethical considerations that might arise in the development and utilisation of digital twins in the built environment, such as data privacy and security issues Deren et al. [4]. It may be imperative to consider these issues individually or develop additional guidelines or frameworks to tackle them.

2.2 *Importance of Maturity Models in Digital Twins*

A maturity model is an indispensable component of creating a digital twin for the built environment. The model serves as a structured framework for evaluating and refining the performance of the digital twin over time. Maturity models, in general, comprise a set of criteria used to assess the current state of any system and identify opportunities for improvement [2]. These criteria encompass aspects such as system integration, automation level, data accuracy, and user engagement. The specific criteria used in a maturity model may vary based on the goals and objectives of the digital twin.

Incorporating a maturity model helps guarantee the consistent progression and satisfaction of stakeholders' needs through the digital twin [7]. It can also assist in determining and prioritising areas for improvement and provide a plan for executing modifications and upgrades [7]. Furthermore, a maturity model can function as a reference point for comparing the digital twin's performance with other similar systems and uncover best practices and successful strategies [7]. Overall, having a maturity model in place can optimise the value and impact of the digital twin and ensure it stays a useful and effective tool for managing and optimising the built environment.

2.3 Importance of Real-Time Data Visualisation Relation to Digital Twins

Real-time data visualisation drives the built environment digital twin adoption [11]. A digital twin can provide a full and dynamic perspective of the campus, helping decision-making and resource optimisation. The digital twin can be a single source for campus data by integrating energy use, traffic patterns, and occupancy data [11, 12]. Data can be easily accessible, analysed, and visualised in one platform, making it more efficient and productive. Visualising data in the digital twin can also reveal patterns, trends, and anomalies not seen in separate data sources.

Campus infrastructure and facilities can be visualised in 3D. A 3D digital twin can be used to analyse campus operations including energy usage, traffic patterns, and tenant behaviour by merging building locations, floor plans, and occupancy rates [9]. It also visualises energy consumption data in connection to building usage patterns can reveal energy conservation potential. Additionally, it visualises traffic patterns helps find bottlenecks and optimise routes [15]. The Geographic Interface System (GIS) can also be used to map campus data [8]. By incorporating data on campus facilities like food and retail outlets, parking spaces, and transportation routes, a dynamic map can be created to analyse user behaviour and foot traffic. Also, GIS data helps indicate regions of the campus heavily used by students and professors, which can help locate future amenities and facilities. Dashboards display real-time campus data while using data from energy use, occupancy sensors, and weather stations. Additionally, our project dashboards may show real-time occupancy rates, and weather conditions and which aim to improve strategies on resource allocation and decision-making.

3 Methodology

In this research, a framework for creating digital twins of university campuses is proposed and evaluated. This was done using a case study methodology, with a university campus as the study's focal point. The St Lucia Campus of the University of Queensland was chosen as the case study for this investigation, owing to its size and intricacy. This campus, situated in Brisbane, Australia, spans an area of approximately 114 ha. and is comprised of over 95 buildings, inclusive of various residential halls, classrooms, administrative structures, and recreational facilities, catering to a diverse range of stakeholders.

Along with a range of teaching and administration focus buildings, the campus boasts a wide range of facilities, such as libraries, athletic fields, and diverse food and commerce options, but it is also the site of numerous research and innovation centres, such as the Queensland Brain Institute and the Australian Institute for Bioengineering and Nanotechnology. This renders the campus an ideal subject for a case study on the creation of a digital twin framework not only for university campuses but also as a small-scale urban environment.

3.1 Creation of a Campus Framework

The creation of a framework for a university campus digital twin presents a unique opportunity to establish a model that can be adapted to a wide range of environments. The complex, multifaceted systems of university campuses—encompassing a variety of spaces from dormitories to lecture halls, administrative offices, and leisure spaces—offer an ideal testing ground for constructing a digital twin framework that can accommodate a wealth of information from multiple sources [18] leading to a deeper understanding of how digital twins can enhance the management and operation of these sprawling systems. Moreover, universities tend to host a rich array of stakeholders such students, professors, staff, and administrators. each with their own unique needs and priorities.

By designing a framework for the digital twin that accommodates this diversity, researchers increase the likelihood of widespread adoption and use by the stakeholders [3]. This leads to a digital twin that is more productive, valuable, and beneficial for the university [6]. Finally, universities are often at the forefront of research and development. By creating a framework for university campus digital twins, researchers can take advantage of this innovative environment to experiment with novel uses of digital twins to improve the management and operation of these complex systems [16, 18].

3.2 Development Blueprint

The creation of an effective digital twin framework for the built environment requires a well-planned and executed approach, to ensure its success and maximise its impact. Guided by the principles of the ISO 23247 standard, the following steps can serve as a blueprint for the development of a digital twin framework (see Fig. 2):

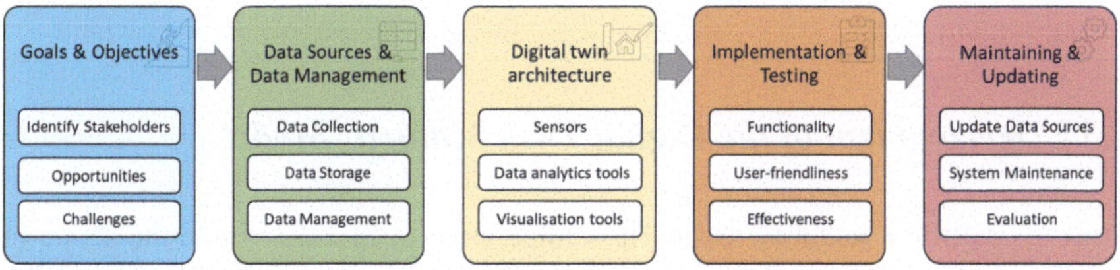

Fig. 2 Diversity of components and categories that might be found in a university campus (*Source* Image created by the authors)

1. Goals and objectives including identifying the different stakeholders and detecting any challenges that might arise and the unique needs of the stakeholders.
2. Scanning the landscape of data sources and establish management and storage protocols to ensure data robustness.
3. Utilisation of sensors, data readers, analytical and visualisation tools as part of the DT architecture of the built environment. It is crucial at that stage to design how various systems will integrate.
4. Implementing and evaluating the digital twin using appropriate methods and tools, such as pilot projects or case studies, to confirm its functionality and effectiveness.
5. Sustaining the digital twin's relevance over time by regularly updating and reviewing the data sources and data management processes, as well as conducting ongoing testing and analysis to pinpoint areas for improvement.

In implementing the modified ISO23247 for the built environment, the initial step is to identify the goals and objectives of the digital twin for the UQ campus. This includes pinpointing the stakeholders, encompassing students, faculty, staff, and visitors, and comprehending their needs and expectations. In addition, it is imperative to consider the opportunities and challenges posed by the digital twin, such as the potential enhancement for efficiency, heightened data-driven decision-making, and improved communication and collaboration among stakeholders. By establishing clear goals and objectives, the digital twin can be effectively synchronized with the needs and goals of the UQ campus community.

One of the main goals of the digital twin for the University of Queensland Saint Lucia Campus is to provide a comprehensive and up-to-date representation of the campus, including its physical spaces, amenities, sensors and meters, services, events and activities, users, and policies and regulations. This digital twin will serve as a tool for understanding the complex systems and processes that operate on the campus and will allow for the identification of opportunities for improvement and optimisation.

Another important objective of the digital twin is to support decision-making and planning processes on the campus. By providing a real-time, data-driven view of the campus, the digital twin will enable stakeholders to make informed decisions about the use of resources and the allocation of space. Additionally, the digital twin will allow for the simulation of various scenarios and the evaluation of their potential impacts, providing valuable insight into the long-term planning and development of the campus.

3.3 Development of the Digital Twin Maturity Model

Prior to crafting a maturity model, it is imperative to first determine the scope of the digital twin. This involves specifying its goals and components it covers. With a clear scope, it is then possible to develop metrics and indicators that are relevant and applicable to the digital twin in question. A few points must be considered when designing the maturity model such as scalability, adaptability to new data sources, user-friendliness, security and privacy, data accuracy and reliability, to ensure that it is relevant and effective across various applications.

The UQ campus maturity model is a tool that aims to assess and evaluate the current state of the university campus in relation to the goals and objectives identified during the planning phase. The model consists of a list of components and categories that constitute the university campus, including physical spaces, amenities, sensors and meters, services, events and activities, users, and policies and regulations. Each of these components and categories is assigned a level of maturity ranging from "Sub-basic" to "Complete", based on their ability to contribute to the achievement of the identified goals and objectives.

The five levels of maturity are defined as follows:

- Sub-basic: Minimal or no capability to contribute to the achievement of the goals and objectives.
- Basic: Limited capability to contribute to the achievement of the goals and objectives.
- Intermediate: Moderate capability to contribute to the achievement of the goals and objectives.
- Advanced: Significant capability to contribute to the achievement of the goals and objectives.
- Complete: Full capability to contribute to the achievement of the goals and objectives.

To illustrate these levels and within the Physical Spaces category, the key performance indicator (KPI) "*3D model*" might be considered "*Sub-basic*" if it only includes basic 3D building models with no data integration, "*Basic*" if it includes 3D models with some data integration, and "*Complete*" if it includes fully detailed 3D models with comprehensive data integration. Similarly, in the infrastructure category, the KPI "Parking spaces" might be considered "*Sub-basic*" if there is no real-time information available on parking availability, "*Basic*" if there is some real-time information available, and "*Complete*" if there is comprehensive real-time information available on parking availability.

In the category of Assets, the key performance indicators "Equipment Monitoring" could be used to assess the level of maturity of the digital twin. At the "*Sub-basic*" level, there may be limited or no monitoring of equipment on campus, with maintenance and repairs being scheduled based on reactive rather than proactive measures. As the digital twin matures, the level of equipment monitoring may increase, with real-time monitoring and analytics being used to predict and prevent equipment failures (see Fig. 3).

At the "Complete Level", the digital twin may include a comprehensive system for monitoring and managing all equipment on campus, including regular maintenance schedules and the use of predictive analytics to optimize equipment performance and extend its lifespan. This KPI is closely aligned with the overall objectives and goals of the digital twin, as it helps to ensure the smooth and efficient operation of the campus and the availability of resources for students, faculty, and staff. Overall, the

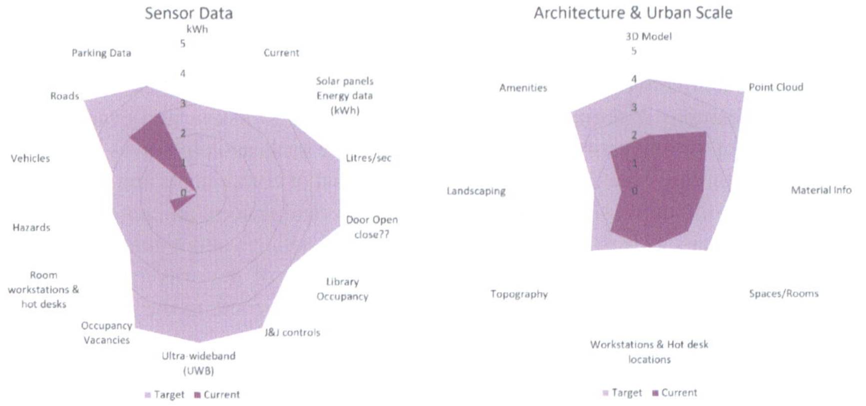

Fig. 3 Maturity Model key performance indicators for the university campus maturity model for architecture, urban and sensor data (*Source* Image created by the authors)

UQ campus maturity model is a useful tool for evaluating the progress and potential of the digital twin, and it helps to identify areas for improvement and development.

3.4 ETL Process and Structure

Method is a widely used process in data management and analysis, and it is particularly useful in the development of digital twins due to its structured and systematic approach. The ETL (extract, transform, load) process is employed as an essential component of the digital twin development process at UQ, as because it allows for the efficient and effective integration of data from various sources and the use of various tools and techniques for data transformation and analysis. By tapping into the ETL method, the UQ campus digital twin can consistently evolve and mature, aligning with the defined goals and objectives outlined in the maturity model. Additionally, the continuous evaluation and improvement provided by the maturity model allows us to adapt and integrate new data sources into the ETL process as they become available. The following diagram illustrates the ETL methods, and the tools and sources used to create the UQ digital twin (see Fig. 4).

The initial stage phase requires involves extracting data to be extracted from various sources, such as spatial modelling software, data acquisition tools, and databases. The second phase, is data transformation, which involves includes the manipulating on and analysis analysing of the collected data using spatial data tools, which in this case used are visual scripting tools like Grasshopper and Dynamo. These tools are geometry-based scripting language or a geometric modelling kernel, which allow enabling data attributes to be associated with geometry and simulation/ analysis plugins or scripts depending tailored on to the intent of the digital twin purposes. This structure allows the generation generates of analysis-ready data, such

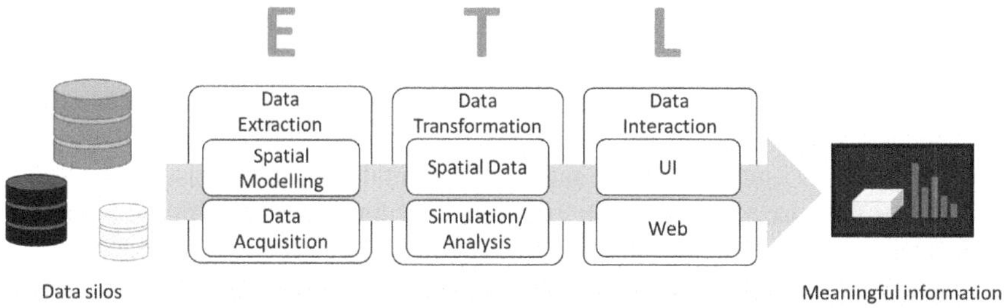

Fig. 4 A general outline of the ETL process used to create the digital twin (*Source* Image created by the authors)

as functional and occupancy maps or flood simulations, which could be used to inform campus planning and policymaking.

Finally, the third phase, contains data interaction, which involves the creation of a user interface (UI) using to a visual interface and point cloud visualisation, as well as the use of web-based interface to make the data accessible and interactive for stakeholders. This step allows for the digital twin to be used for real-time decision-making and for stakeholders to explore the data in a more intuitive and interactive manner [9].

The ETL method was used in the development of a digital twin for the university campus, enabling the incorporation of data from various sources and the use of various tools and techniques for data transformation and analysis. This structured and systematic approach facilitated the creation of a comprehensive and useful digital twin of the campus, demonstrating the efficiency and effectiveness of the ETL method.

Additionally, the ETL method allows for the incorporation of a wide range of data inputs and can be integrated across all phases of the digital twin development process. This versatility and flexibility make it an effective approach for creating comprehensive and useful digital twins in various settings.

The framework considers the various data sources and tools needed to create a comprehensive and dynamic digital twin, as well as the potential benefits and challenges of using digital twins in the management of university campuses Vivi et al. [16]. In addition, a discussion is presented on the potential applications of the framework for other large-scale digital twin settings such as towns and cities Kim et al. [7].

3.5 ETL University Campus Digital Twin Custom Framework

In the case study for the University of Queensland Saint Lucia Campus the ETL framework from above is used for the incorporation of various data inputs and integration of input data across all phases of digital twin development. The UQ St. Lucia campus Digital Twins creation requires a matrix of flexible data type inputs which requires and alignment with an expanded ETL framework (see Fig. 5).

In the data acquisition phase, six main data inputs were employed: OpenStreetMap data of the campus, Archibus data, UQ Properties and Facilities data, live data, point cloud data and 2D satellite images, and environmental data. In the case study, Rhino was used as the base to model, transform and load the data for the digital twin with the help of the visual scripting tool Grasshopper.

Regarding data transformation phase, the data inputs were cleaned and processed according to their specific data types. Data processing included various simulations and predictions using tools such as the Elk plugin for sorting OpenStreetMap data, the Ladybug plugin for environmental simulations, native Grasshopper components

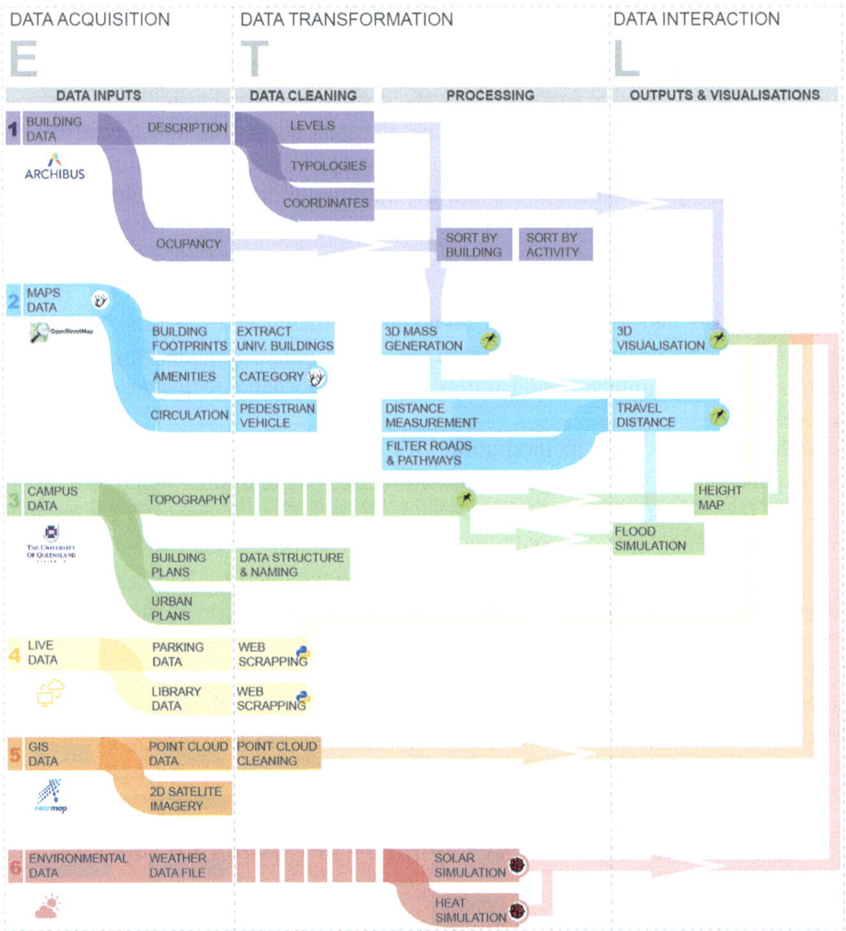

Fig. 5 The ETL process for creating DTw expanded to accommodate flexible data type inputs for the UQ campus (*Source* Image created by the authors)

for geometry and topography generation, custom Python scripts for web scraping and other functions, and the Volvox and Tarsier plugins for point cloud data sorting and visualisation.

The processed and cleaned data were then combined and presented as various outputs, including a 3D visualisation of the entire campus showing different layers of data sources such as 3D building masses, height maps, flood simulation results, solar radiation results, and energy consumption results. Finally, in the data interaction phase, a user interface was created using the Human UI plugin within Grasshopper.

This workflow demonstrated the flexibility of the ETL method in integrating new data inputs and employ a range of tools and techniques for data transformation and analysis. It allowed for the creation of a comprehensive and useful digital twin of the university campus, highlighting the efficiency and effectiveness of the ETL method.

3.6 Campus Level Data

At the core of our Digital Twin framework is the Campus Level Data interface, a visualization tool that integrates detailed campus maps, building information, and infrastructure details. This interface not only maps physical structures through a combination of 3d modelling and point clouds but also overlays additional data layers such as energy consumption metrics, environmental conditions, and accessibility features, providing a comprehensive view of campus dynamics. Key elements like building heritage and indigenous knowledge are prominently featured, enriching the campus narrative, and contributing to a richer cultural appreciation. This distinct array of information will be unpacked through four main sets: Dynamic Historical Visualization, Functional Diversity and Usage Analysis, Building Level Data and Environmental and Mobility Simulations.

3.7 Visualisation and Interface Capabilities

Central to our Digital Twin framework is the Campus Level Data interface, a sophisticated visualisation tool that integrates detailed campus maps, building information, and infrastructure details (Fig. 6). This interface not only maps physical structures but also overlays additional data layers such as energy consumption metrics, environmental conditions, and accessibility features, providing a comprehensive view of campus dynamics. Key elements like building heritage and indigenous knowledge are prominently featured, enriching the campus narrative, and contributing to a richer cultural appreciation.

One of the standout features of our digital twin is its capacity for dynamic historical visualisation (Fig. 7). This allows stakeholders to view the evolution of the campus over time—tracking architectural growth, changes in usage, and the expansion of

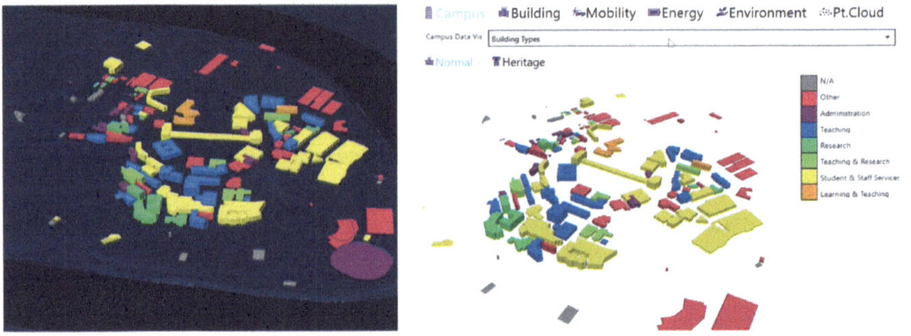

Fig. 6 Interface showing the different information that can be accessed and visualised in UQ campus level data interface (*Source* Image created by the authors)

facilities. This temporal perspective is invaluable for planning and demonstrates the adaptability of campus infrastructure to evolving academic and community needs.

Functionality and usage data are color-coded within the digital twin, providing intuitive insights into building use through wireless monitoring and counters located building entrances and carparks. This helps campus managers and policymakers understand and improve how space is used through IoT approach, enhancing efficiency, and ensuring that facilities align with the needs of the campus community [8].

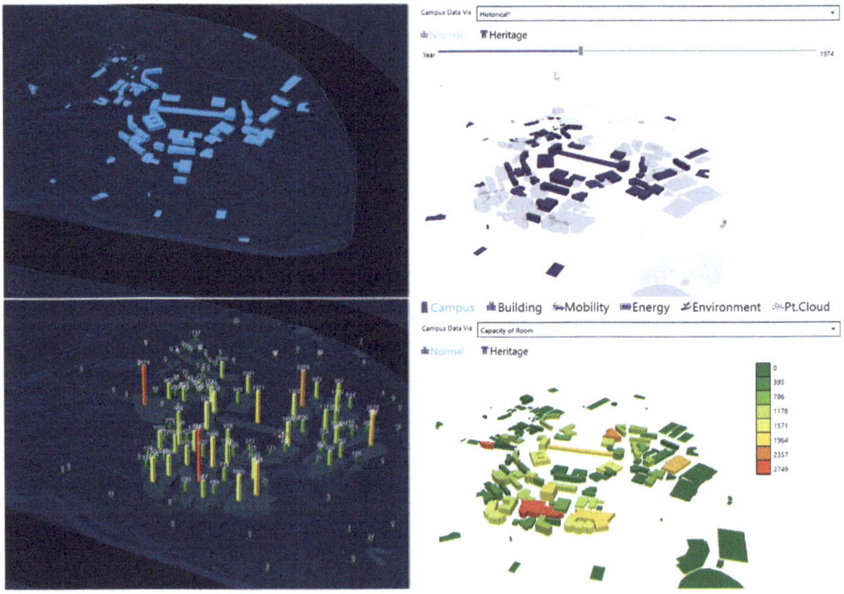

Fig. 7 Interface showing the historical visualisation of building dates and campus dynamic growth (top) and buildings occupancy capacity (bottom) (*Source* Image created by the authors)

Live and historical occupancy data further refine this analysis, offering a granular look at campus life and activity patterns.

3.8 Building Level Data

In addition to broader campus insights, our digital twin drills down to detailed building-level data. Each building's digital representation includes comprehensive information such as localisation, architectural history, and current usage, supplemented by BIM models and point cloud data. This granularity supports maintenance and future development plans, ensuring that interventions are well-informed and contextually appropriate (see Fig. 8).

The Building Level Data is further enhanced by the ability to add point cloud data, either from a satellite source or acquired through 3d scans that are regularly produced as a form of mapping building updates (see Fig. 9).

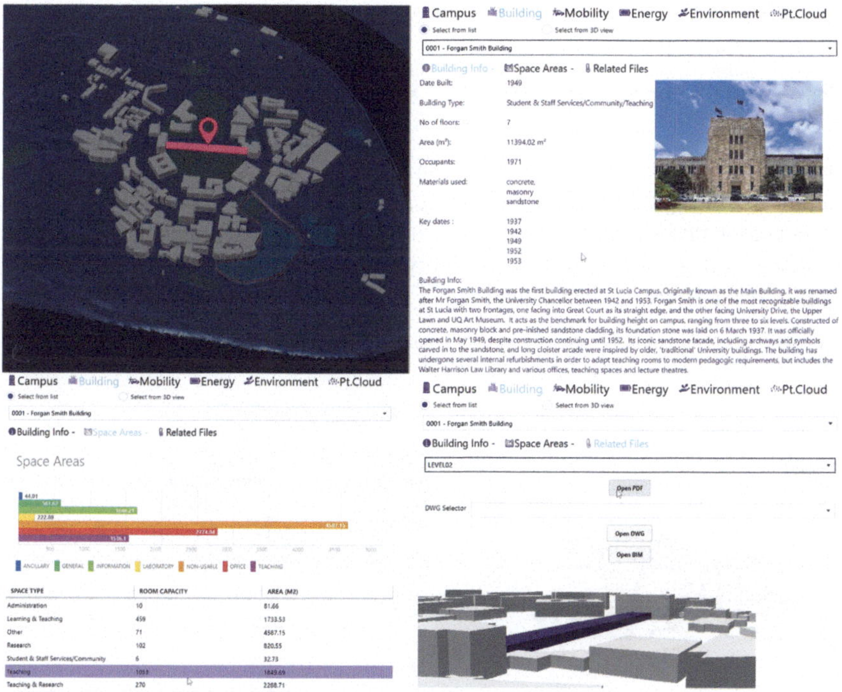

Fig. 8 Screenshot showing the localisation, history, typology, and ability to download any additional visual resources (*Source* Image created by the authors)

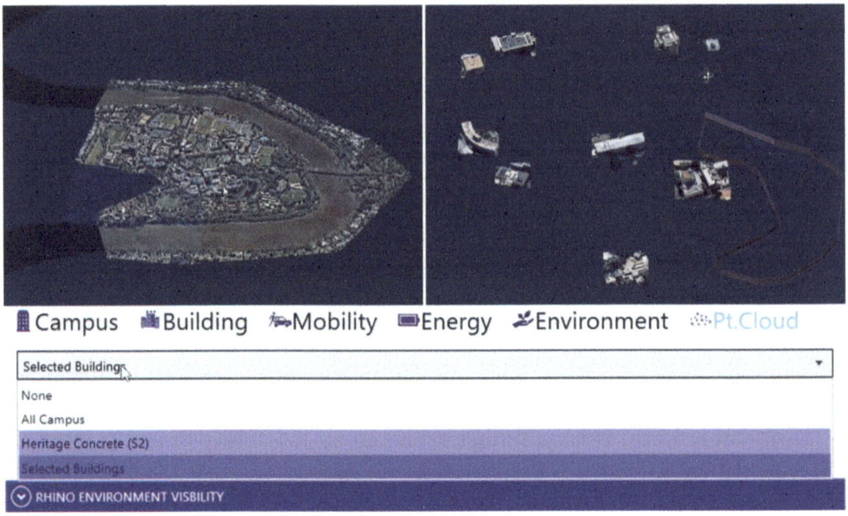

Fig. 9 Screenshot showing the satellite point cloud (left) and building specific point cloud information (right) (*Source* Image created by the authors)

3.9 *Environmental and Mobility Simulations*

Advanced simulations of environmental impacts and mobility patterns form a critical part of our digital twin. By modelling scenarios such as solar radiation impact and potential flood zones, we can proactively address environmental challenges and sustainability goals. Similarly, mobility simulations help optimise traffic flows and pedestrian pathways, enhancing safety and accessibility across the campus (see Fig. 10).

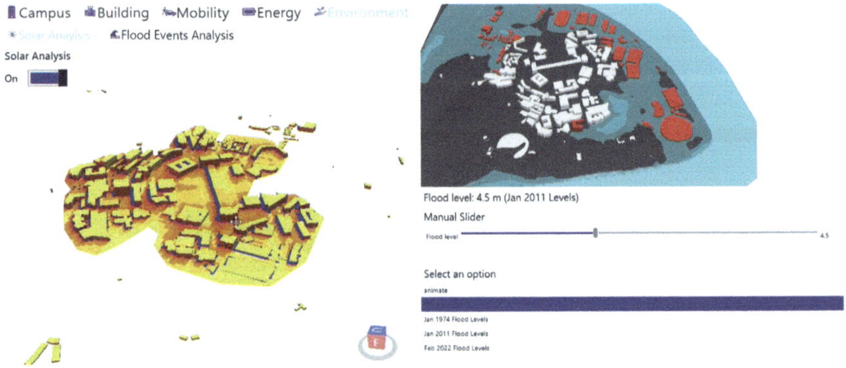

Fig. 10 Interface showing solar analysis (left) and flood analysis (right) (*Source* Image created by the authors)

Under the mobility simulation mobility patterns through the combined use of geographical information such as roads and walkways. This is applicable to both pedestrian and car traffic which allows to predict busiest routes as well as alternative routes in case of disruption and difference models (see Fig. 11). In addition, due to the spatial information embed in the walkway we can also have improved, or alternative routes based on universal access parameters.

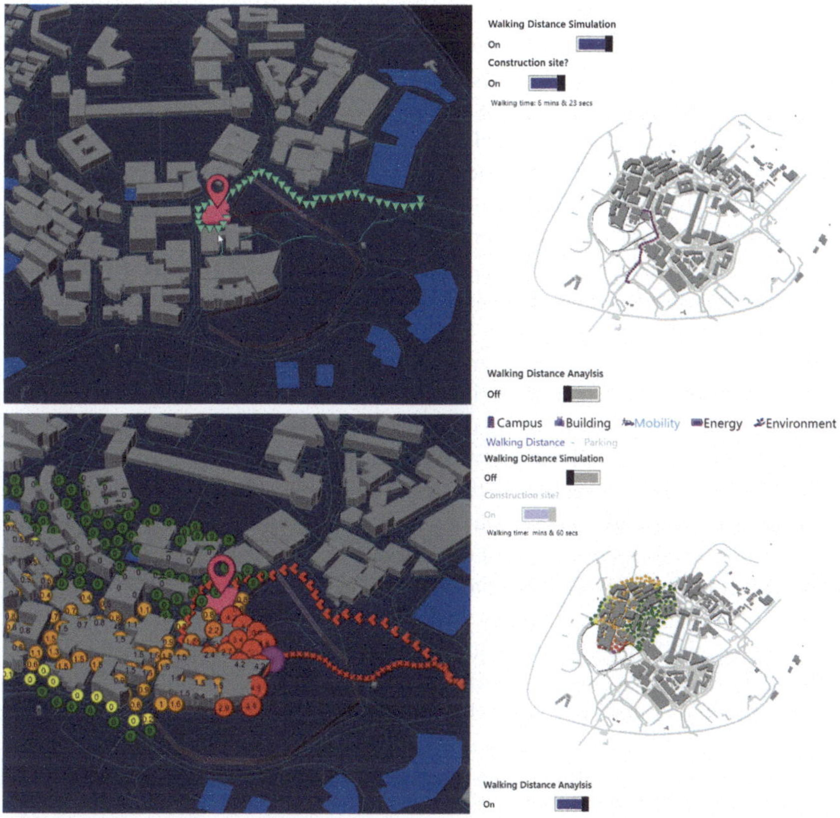

Fig. 11 Screenshot presenting walking simulations (top) and delayed arrival times (bottom) due to detour (*Source* Image created by the authors)

4 Discussion and Implications

The integration of these diverse data layers allows for sophisticated interpolation and analysis. By correlating various datasets—such as energy usage with occupancy rates—we can identify inefficiencies and optimise operations. This analytical capability is pivotal in transitioning the campus towards greater sustainability and operational excellence. Certainly, even within a large entity there are challenges in data collection, such as discrepancies in data accuracy or delays in data transmission. Nonetheless, solutions emerge, and these include do include the clear exchange and implementation of a data governance framework to uphold data integrity and the deployment of advanced data processing tools to handle large datasets effectively.

Spatial information and advanced analytical techniques play a significant role in data interpolation and analysis, supporting in the identification of patterns and inefficiencies. The example utilises the Campus Level Data and 3D visualisation tools to represent campus infrastructure and facilities. By integrating data on building locations, floor plans, and occupancy rates into a 3D digital twin, it is paramount to create a spatial representation of the campus that can be used to analyse various aspects of campus operations, such as energy consumption, traffic patterns, and occupant behaviour. For example, energy consumption data can be visualised in relation to building usage patterns to identify opportunities for energy conservation. Similarly, traffic patterns can be visualised to identify bottlenecks and optimised transportation routes and all using spatial representation.

The UQ digital twin developed in this research enabled the visualisation of historical data in a meaningful way, providing valuable insights into the past performance and usage of the university campus. For example, by incorporating data on the dates that buildings were constructed and the historical occupancy rates of those buildings, the digital twin allowed for a deeper understanding of the evolution and use of the campus over time. Similarly, the inclusion of data on previous flood events allowed for the identification of potential vulnerabilities and the development of strategies for mitigating future flood risk [13]. The ability to use data interpolation and analysis within a spatial interface of the digital twin framework enhances the understanding of the campus and its operations, informing decision-making and planning processes [6].

Overall, the UQ Campus Digital Twin framework highlights potential applications of integrated spatial data within a Digital Twin framework. Indeed, the "Campus Level Data" section underscores the potential of digital twins in transforming university campuses into smarter, more sustainable environments. Achieving this transformation requires thorough data integration and advanced spatial analytics. Our framework not only enhances current management practices but also paves the way for future innovations in campus planning and development. This approach ensures that the campus remains adaptable and responsive to the evolving needs of its community, setting a benchmark for digital twins in university campuses and the built environment at large. The research findings and key lessons of the UQ Digital Twin Campus can be itemised as follows:

- Enhanced Data Interpolation and Analysis: Integration of diverse data layers enables sophisticated data interpolation and analysis, crucial for identifying inefficiencies and optimizing campus operations.
- Data Governance and Processing: Effective data governance frameworks and advanced data processing tools are essential for maintaining data integrity and handling large datasets.
- Spatial Representation: Utilizing 3D digital twins for spatial representation of campus infrastructure aids in analysing various operational aspects, including energy consumption and traffic patterns.
- Historical Data Insights: Visualizing historical data within the digital twin provides valuable insights into campus evolution and performance, aiding in risk assessment and strategic planning.
- Sustainable Campus Operations: The digital twin framework supports the identification of opportunities for energy conservation and optimization of transportation routes, contributing to a smarter, more sustainable campus.
- Adaptive and Responsive Planning: The framework ensures that the campus remains adaptable to evolving needs, setting a benchmark for digital twin applications in university campuses and the built environment.

These findings highlight the transformative potential of digital twins in enhancing campus sustainability and operational excellence through the combined use of advanced data integration and spatial analytics.

5 Conclusion

This research developed a comprehensive framework for the integration of digital twins into university campus management, with potential applicability extending to urban environments. Through the employment of advanced data integration and visualisation techniques, this framework addresses the pressing need for cohesive and scalable digital twin technologies that can significantly enhance urban sustainability efforts. The case study of a university campus digital twin demonstrates the framework's ability to harness underutilised data, thus providing a robust basis for simulations and spatial analysis essential for informed decision-making in campus planning and development. One clear advantage of the Digital Twin framework for University Campus Applications developed using readily available tools such as Rhino, Grasshopper, OpenStreetMaps, and Archibus, along with point cloud models. It integrates various data sources, including energy consumption, traffic patterns, and occupancy rates, which are visualised to potentially improve various aspects of campus operations. This is an economical and highly effective tool by built environment specialists for built environment specialists.

The ability to synthesise various data layers into a single spatial information repository, the framework not only optimises the management of physical assets but also enhances the strategic planning capabilities of urban settings. The integration process involved combining and interpolating data layers to create a cohesive digital twin environment. Challenges and limitations encountered during this process were addressed to ensure the reliability and validity of the results, although it is noted that the study's scope being limited to a single case study may introduce potential bias or errors.

The implications of this research are far-reaching, offering urban planners and architects a novel approach to incorporate sustainability into the lifecycle management of urban infrastructures. The predictive capabilities of digital twins, as outlined in this study, facilitate a proactive stance in urban management—anticipating problems, mitigating risks, reducing costs, and improving overall operational efficiency.

Future research should explore the scalability of this framework to larger urban contexts, assess its adaptability to different environmental and operational parameters, and refine its predictive analytics features. Such actions will expand the utility of digital twins, making them a cornerstone technology in the pursuit of sustainable urban development. It is hoped that the insights derived from this framework will inform future research and practice in the area, guiding the development of more sophisticated digital twins for the built environment.

Acknowledgements We extend our gratitude to Kelton Boyter-Grant, Ben Ziegenfusz, and Christian Dorst for their invaluable contributions to the project. Additionally, we acknowledge the commitment and support of UQ Properties and Facilities, as well as the UQ Library. Their collaboration has been instrumental in advancing our research and enhancing the outcomes of this project.

References

1. Alonso R, Borras M, Koppelaar R H E M, Lodigiani A, Loscos E, Yöntem E (2019) SPHERE: BIM digital twin platform 9. https://doi.org/10.3390/proceedings2019020009
2. Auditors, TI of I (2018) Selecting, using and creating maturity models: A tool for assurance and consulting engagements. The institute of internal auditors. https://www.theiia.org/en/content/guidance/recommended/supplemental/practice-guides/selecting-using-and-creating-maturity-models/
3. Broo DG, Schooling J (2021) Digital twins in infrastructure: definitions, current practices, challenges and strategies. Int J Constr Manag 23(7):1254–1263. https://doi.org/10.1080/15623599.2021.1966980
4. Deren L, Wenbo Y, Zhenfeng S (2021) Smart city based on digital twins. Comput Urban Sci 1(1):1–11. https://doi.org/10.1007/S43762-021-00005-Y/FIGURES/12
5. Hakimi O, Liu H, Abudayyeh O (2023) Digital twin-enabled smart facility management: a bibliometric review. Front Eng Manage 2023:1–18. https://doi.org/10.1007/S42524-023-0254-4
6. Han X, Yu H, You W, Huang C, Tan B, Zhou X, Xiong N N (2022). Intelligent campus system design based on digital twin. Electronics 11(21):3437. https://doi.org/10.3390/ELECTRONICS11213437
7. Kim Y-W, Yoo S, Lee H, Kim S, Han S (2020) Characterization of digital twin

8. Li Z, M L, C W, Li F (2020) Design and application of intelligent monitoring and management system for campus weak current pipeline network based on IoT+ GIS technology. J Shenzhen Univ Sci Eng 37(S1):155–159
9. Meng L, Ji K, Wang J (2021) Development of web-based virtual campus roaming system. Comput Eng Des
10. Oliveira PP (2020) Digital twin development for airport management. J Airpt Manag 14(3):246–259
11. Opoku DGJ, Perera S, Osei-Kyei R, Rashidi M, Famakinwa T, Bamdad K (2022) Drivers for digital twin adoption in the construction industry: A systematic literature review. Buildings 12(2). MDPI. https://doi.org/10.3390/buildings12020113
12. Ozturk GB (2021) Digital twin research in the AECO-FM industry. J Build Eng 40:102730. https://doi.org/10.1016/J.JOBE.2021.102730
13. Shahzad M, Shafiq MT, Douglas D, Kassem M (2022) Digital twins in built environments: an investigation of the characteristics, applications, and challenges. Buildings 12(2). https://doi.org/10.3390/buildings12020120
14. Shao G, Helu M (2020) Framework for a digital twin in manufacturing: scope and requirements. Manuf Lett 24:105–107. https://doi.org/10.1016/j.mfglet.2020.04.004
15. Sleiti AK, Kapat JS, Vesely L (2022) Digital twin in energy industry: proposed robust digital twin for power plant and other complex capital-intensive large engineering systems. Energy Rep 8:3704–3726. https://doi.org/10.1016/J.EGYR.2022.02.305
16. Vivi QL, Parlikad AK, Woodall P, Ranasinghe GD, Heaton J (2019) Developing a dynamic digital twin at a building level: using Cambridge campus as case study. In: International conference on smart infrastructure and construction 2019, ICSIC 2019: driving data-informed decision-making, pp 67–75. https://doi.org/10.1680/ICSIC.64669.067
17. Wang W, Guo H, Li X, Tang S, Xia J, Lv Z (2022) Deep learning for assessment of environmental satisfaction using BIM big data in energy efficient building digital twins. Sustain Energy Technol Assess 50. https://doi.org/10.1016/j.seta.2021.101897
18. Zaballos A, Briones A, Massa A, Centelles P, Caballero V (2020) A smart campus' digital twin for sustainable comfort monitoring. Sustainability (Switzerland) 12(21):1–33. https://doi.org/10.3390/SU12219196

Roads Infrastructure Digital Twin Case Study: Uzbekistan

Jamshid Sodikov

Abstract The integration of Digital Twins (DTs) into road infrastructure management presents a ground breaking approach to addressing the multifaceted challenges of modern transportation networks. The study explores the evolution of DTs from their inception to their current role as sophisticated decision-support systems that incorporate real-time data, artificial intelligence (AI), and geographical information systems (GIS) to enhance infrastructure management. Focusing on Uzbekistan, the research outlines the potential of DTs to revolutionize road maintenance, traffic management, and environmental resilience in facing geographical and climatic extremities. This paper presents a detailed case study on the implementation feasibility, strategic benefits, and socio-economic impacts of DTs in improving road infrastructure management in Uzbekistan. By integrating cutting-edge technologies and data analytics, this study highlights the transformative potential of DTs to foster predictive maintenance strategies, optimize resource allocation, and improve road safety and sustainability. The research methodology encompasses an extensive review of current practices, data collection strategies, and the development of a digital twin system tailored to the challenges and opportunities within the Uzbek road sector. The findings underscore the significance of DTs in enhancing decision-making processes, reducing maintenance costs, and contributing to sustainable urban development. The study provides a comprehensive framework for the deployment of digital twins, including considerations for data privacy, technological integration, and stakeholder engagement. Furthermore, it offers policy recommendations for the adoption of DTs as a strategic tool in urban planning and infrastructure management, advocating for a collaborative approach to innovation and capacity building. The paper contributes valuable insights into the role of digital twins in modernizing road infrastructure management. It delineates a path forward for leveraging advanced technologies to create more resilient, efficient, and intelligent transportation systems, with broader implications for global infrastructure management practices.

J. Sodikov (✉)
Urban Planning Department, Tashkent State Transport University, Tashkent, Uzbekistan
e-mail: sodiqov_j@tstu.uz

S. Pourroostaei Ardakani and A. Cheshmehzangi (eds.), *Digital Twin Computing for Urban Intelligence*, Urban Sustainability,
https://doi.org/10.1007/978-981-97-8483-7_10

1 Introduction

The concept of the digital twin has its roots in NASA's Apollo program, which involved the construction of multiple identical spacecraft. This methodology was crucial for simulating the conditions of the spacecraft in space while on a mission. Building upon this foundation, Shafto et al. [1] later defined the NASA digital twin as an encompassing simulation that integrates various physical dimensions and scales. This simulation, which incorporates the best available physical models, sensor data, and historical information about the fleet, serves to accurately reflect the life cycle of the spacecraft it represents [1]. The concept of Digital Twins (DTs) has emerged as a transformative force in the realm of infrastructure management, marking a new era in the application of advanced technologies. Originating in 2002, DTs have evolved from simple digital mirrors of physical objects to complex decision-support tools integral to modern engineering and urban planning [2, 3]. This evolution reflects a broader shift toward data-driven, predictive, and intelligent management strategies across various sectors, including manufacturing, urban development, and transportation [4]. In urban planning and city infrastructure management, DTs have garnered attention for their potential to leverage vast urban data pools. By harnessing this data, urban planners and city infrastructure managers can significantly enhance decision-making processes, supported by new-generation IT tools such as Artificial Intelligence and Big Data Analytics [5]. This paradigm shift is not just about data collection but encompasses a comprehensive approach to infrastructure management, including monitoring, analysis, prediction, and proactive intervention. DTs in the road sector, in particular, offer a multidimensional and dynamic approach to infrastructure management. They provide a platform for continuous and detailed monitoring of road conditions, enabling a shift from reactive to predictive maintenance strategies. This approach is crucial for managing the complexities of road networks, including traffic flow, pavement conditions, and environmental impacts [6]. The implementation of DTs in road infrastructure involves multiple layers of data collection and analysis. The physical layer, encompassing technologies such as Lidar Scanners, RGB cameras, and UAVs, collects comprehensive data throughout a road's lifecycle. This data, encompassing various road assets and conditions, feeds into an asset management database, enhancing both operational and maintenance strategies [7]. The integration of such technologies facilitates a nuanced understanding of road behavior under varying conditions, informing better infrastructure planning and maintenance. The value of DTs extends to dynamic road maintenance and management, where continuous streams of data from diverse sources enable real-time adjustments and predictive maintenance. Such a system allows road owners and managers to promptly respond to changes in traffic levels, emissions, or pavement conditions, thereby enhancing the resilience and efficiency of road networks [6]. The ability to integrate and analyze data from multiple sources in real-time represents a significant advancement over traditional snapshot-based approaches to road management.

As we focus on Uzbekistan, a country with unique geographical and climatic challenges, the implementation of DTs in road infrastructure presents both opportunities and challenges. Uzbekistan's road network, a crucial component of its economic and social fabric, faces pressures from climatic extremes and rapid urbanization. The application of DTs in this context offers a pathway to more resilient, efficient, and intelligent road infrastructure management. Our focus in this paper is on the emerging use of DTs in the context of Uzbekistan's road infrastructure. We examine not only the technical feasibility but also the strategic integration of DTs with the country's specific environmental, technological, and socio-economic landscape. Environmental Considerations: Uzbekistan, characterized by its varied climatic conditions, faces challenges such as extreme temperature fluctuations and potentially harsh weather events. These environmental factors critically impact road durability and maintenance needs. The application of DTs in this context is poised to offer a dynamic and responsive approach to monitoring and managing these environmental impacts on the country's road infrastructure. By continuously collecting and analyzing environmental data, DTs can help in predicting and mitigating the effects of weather changes on road conditions, thereby enhancing road safety and longevity. Technological Integration: Uzbekistan's journey towards incorporating advanced technologies in infrastructure management presents both opportunities and challenges. The integration of DTs involves harnessing cutting-edge technologies like IoT, AI, and machine learning to create a comprehensive and real-time understanding of the road network. This integration aims to transform the traditional road management practices into a more proactive and predictive model, ensuring efficient use of resources and improved decision-making. Socio-Economic Impacts: The road network in Uzbekistan is a critical component of its economic development and social welfare. Efficient and well-maintained roads are essential for trade, access to education, healthcare, and other services. The implementation of DTs in road infrastructure management could lead to significant improvements in transportation efficiency, reduce maintenance costs, and potentially stimulate economic growth. Moreover, this technological advancement aligns with the broader socio-economic goals of enhancing quality of life and fostering sustainable urban development in Uzbekistan. Through the lens of these environmental, technological, and socio-economic perspectives, the paper aims to provide a comprehensive analysis of how DTs can be effectively implemented in Uzbekistan's road infrastructure. This discussion will encompass current challenges, potential solutions, and strategies for future development, ultimately presenting a nuanced view of the role of DTs in shaping the future of road infrastructure management in Uzbekistan.

The primary aim of this chapter is to explore the feasibility and potential impact of implementing DTs in the road infrastructure management of Uzbekistan. The specific objectives of the study are:

- To analyze the current state of road infrastructure management in Uzbekistan and identify key challenges and opportunities for improvement.
- To develop a comprehensive framework for the integration of DTs in road infrastructure management, incorporating real-time data, AI, and GIS technologies.
- To assess the socio-economic impacts of DT implementation on road maintenance, traffic management, and environmental resilience.

- To provide policy recommendations and strategic guidelines for the adoption of DTs in Uzbekistan's road infrastructure sector.

2 Literature Review

Within the vast expanse of academic research, platforms like Semantic Scholar stand out, serving as repositories for an expansive collection of scholarly papers. Encompassing over 200 million papers from diverse scientific disciplines, Semantic Scholar not only facilitates access to a broad spectrum of knowledge but also provides a rich dataset that fuels the development of other AI-driven analytical tools. A query for the term "road digital twin" in Semantic Scholar reveals an impressive corpus of research, exceeding 36,000 papers spanning from 1934 to 2024. A visual examination of Fig. 1 illustrates a notable trajectory in publication volume. Over the last decade, there has been a steady uptick in the output of research in this domain. However, the last five years have witnessed a dramatic proliferation, culminating in an annual publication rate exceeding 6,000 papers. This surge underscores the escalating academic and industry interest in the digital twin technology's application to road infrastructure, reflecting broader trends in smart city initiatives and the evolution of Industry 4.0 principles. Such trends indicate a collective recognition of the potential that digital twins hold in revolutionising the design, maintenance, and operation of road networks—pivotal components of modern transport systems.

The burgeoning corpus of research on road digital twins within the past half-decade can be ascribed to a confluence of pivotal factors:

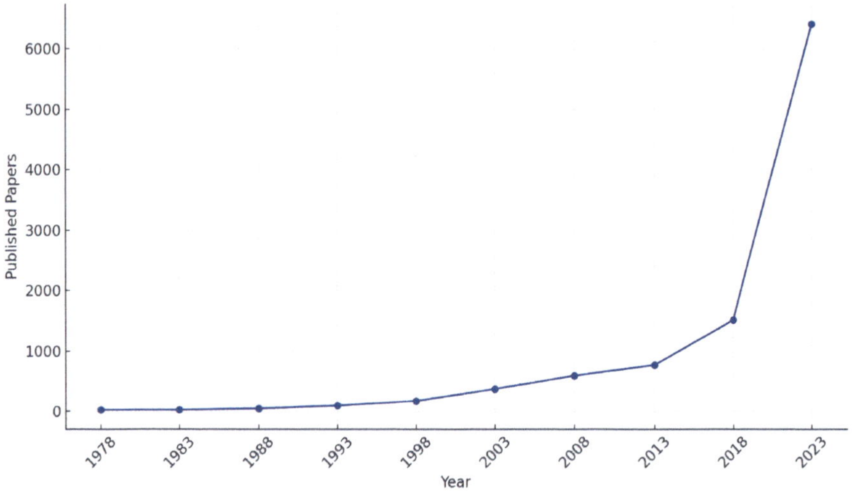

Fig. 1 Number of published paper

- Technological Evolution: The landscape of foundational technologies underpinning digital twins has undergone a profound transformation. The Internet of Things (IoT), artificial intelligence (AI), machine learning, big data analytics, and cloud computing have all reached maturity, ushering in an era of enhanced sophistication and scalability in digital twin applications. This maturation has precipitated the integration of complex digital twin frameworks into mainstream use.
- Data Proliferation: The advent of advanced sensor technology and innovative data collection methods has democratized data availability. The cost-effectiveness of these technologies has precipitated a deluge of data, which in turn fuels the precision and refinement of digital twin models.
- Urban Innovation Imperatives: As cities across the globe vie for smarter, more sustainable futures, digital twins have emerged as central to the urban innovation narrative. This technology is pivotal in managing urban infrastructure, prompting both governmental and private sector investment towards crafting efficient, sustainable, and robust city environments.
- Fourth Industrial Revolution: The principles of Industry 4.0, hallmarked by automation and seamless data exchange, have permeated the infrastructure sector. This shift has been embraced by industry practitioners who recognize the potential of digital twins to enhance the entire lifecycle of road infrastructure, from its inception to the twilight of its use.
- Collaborative Synergy: The intersection of academic inquiry and industry pragmatism has borne fruitful collaborations. These partnerships are dedicated to unraveling the complexities of infrastructure challenges through the lens of digital twin technology, yielding a proliferation of research and innovation.
- Policy and Governance Insights: Policymakers and regulatory entities are increasingly cognizant of the strategic value that digital twins represent for infrastructure stewardship. This awareness has catalyzed a trend towards research funding specifically earmarked for the advancement of these technologies.
- Environmental Imperatives: In the shadow of escalating climate change ramifications, the call for resilient infrastructure has never been more urgent. Digital twins offer a predictive model to navigate and mitigate the impact of environmental flux on road infrastructure, advocating a proactive stance over a traditionally reactive one.
- Investment Trends: The infusion of capital into digital twin technology from both public and private sectors is reflective of a wider commitment to digital transformation agendas. This financial backing is instrumental in driving forward the frontiers of research and the application of digital twins.

These multi-dimensional drivers coalesce to account for the exponential growth in research dedicated to road digital twins, signaling a robust, interdisciplinary interest in this cutting-edge domain.

Recent applications of Digital Twin technology in transportation infrastructure, including railways, highways, bridges, and tunnels, have proven effective in addressing design, construction, and maintenance challenges by Gao [8] highlights benefits like improved real-time state assessment and predictive capabilities, essential for infrastructure integrity. The study identifies key research gaps: enhancing real-time monitoring and prediction, managing the increasing data volume from equipment monitoring, and optimizing design and construction processes. Addressing these gaps will contribute to the optimization and sustainability of transportation infrastructure using Digital Twin technology.

Marai [9] presents a Digital Twin Box (DTB) platform for creating digital twins of roads using real-time data from IoT devices and cameras. The platform addresses bandwidth issues by locally processing video streams and transmitting only the results of object detection and recognition. Real-world experiments demonstrate the effectiveness of this approach, which is crucial for self-driving vehicles and smart mobility. The study highlights the importance of digital twins in enhancing national security through object detection and recognition, aiding in tracking suspicious persons and vehicles and reducing crimes like car theft. Additionally, it discusses the shift towards automated systems and the benefits of interconnectivity in improving safety, efficiency, productivity, energy consumption, environmental protection, and sustainability, aligning with the vision of smart cities and a better quality of life.

Chen [10] conducted a systematic literature review to identify suitable machine learning approaches for road digital twins in road condition predictions. The study highlights AI algorithms such as Support Vector Machines (SVM) for their effectiveness in classification and regression tasks, making them suitable for predicting road conditions based on historical data. Random Forest and Decision Trees are noted for handling large datasets and providing insights into feature importance, aiding road asset management and condition assessment. Neural Networks, with their deep learning capabilities, offer complex pattern recognition and predictive modeling, beneficial for future condition predictions in road digital twins.

Wu [11] finds that Digital Twins (DTs) and Artificial Intelligence (AI) can significantly improve transportation infrastructure in smart cities. DTs can create digital replicas of transportation networks, enabling the simulation and testing of different scenarios to enhance efficiency and safety. AI can analyze data from traffic sensors and cameras to identify patterns and trends, aiding in the development of effective traffic management strategies. The study highlights the potential of DTs and AI to revolutionize transportation through intelligent accident detection, improved traffic flow management, and the integration of new media.

Ashtarout [12] highlights the evolution of state Departments of Transportation (DOTs) in managing transportation assets through data-driven decision-making. The paper emphasizes challenges in cross-asset system analysis and data integration, underscoring the need for Digital Twins in infrastructure. It defines Digital Twins within transportation asset management and examines state DOTs' practices in digitizing asset data, focusing on systems like roadside, electronic, and drainage. The study also explores the timeline for digitization and factors such as data format, detail, and collection techniques, providing insights into the maturity of state DOTs in advancing digital asset management.

Chen [13] presents a pipeline for creating digital twins of off-road environments, focusing on modeling vegetation and uneven terrain. LiDAR scans and scan registration algorithms are used to reconstruct a point cloud map of the environment. Techniques like terrain segmentation, vegetation segmentation, and Euclidean clustering separate point cloud objects into individual entities within the digital twin model. Experimental validation was conducted using LiDAR scans from an off-road proving ground at the Center of Advanced Vehicular Systems (CAVS) at Mississippi State University. The framework, demonstrated with the Mississippi State University Autonomous Vehicle Simulator (MAVS), shows potential applications in virtual autonomous vehicle testing, synthetic data generation, and AI model training.

3 Proposed Digital Twin System for Smart Road Infrastructure Framework

The concept of a Digital Twin (DT) for road infrastructure is built upon the need for an advanced, data-driven approach to managing and maintaining road networks. Traditional methods of road management often rely on periodic inspections and reactive maintenance, which can lead to inefficiencies, higher costs, and reduced road safety. In contrast, a DT system leverages real-time data, advanced analytics, and predictive modeling to create a dynamic, accurate representation of the physical infrastructure. This allows for proactive and informed decision-making.

The motivation for proposing a DT system in Uzbekistan's road infrastructure stems from several key considerations:

1. Environmental Challenges: Uzbekistan experiences extreme temperature fluctuations and harsh weather conditions that significantly impact road durability and maintenance needs. A DT system can continuously monitor these environmental factors and predict their effects on road conditions, enabling timely interventions.
2. Technological Advancements: The integration of IoT, AI, and machine learning technologies provides an unprecedented opportunity to collect and analyze vast amounts of data. This technological capability supports the creation of a comprehensive, real-time understanding of the road network.
3. Socio-Economic Benefits: Efficient and well-maintained roads are vital for economic development and social welfare. By improving transportation efficiency and reducing maintenance costs, a DT system can contribute to economic growth and enhance the quality of life.
4. Data-Driven Decision Making: The ability to integrate and analyze data from multiple sources in real-time allows for more accurate and timely decision-making. This can significantly improve resource allocation, maintenance planning, and traffic management.

The figure 2 visually represents the architecture of a digital twin system for road infrastructure, integrating various data sources, processing mechanisms,

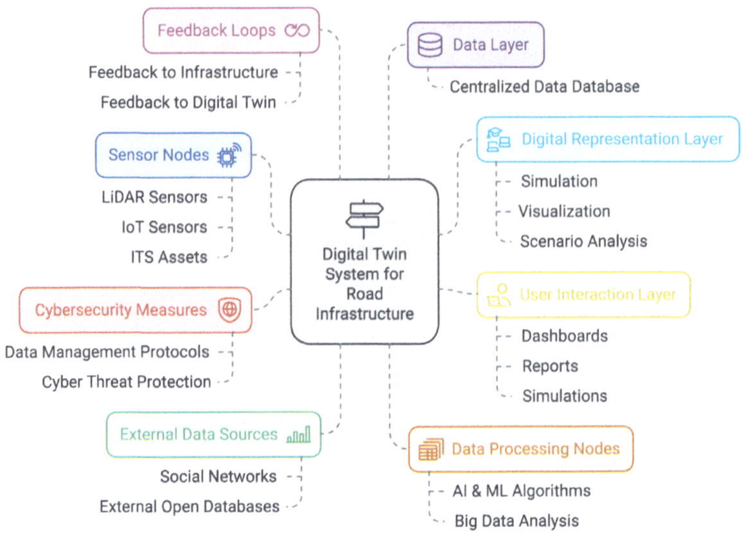

Fig. 2 Integrated digital twin system for smart road infrastructure

and feedback loops. Here's a detailed description and the rationale behind each component:

- Cybersecurity Measures and Data Management Protocols: These elements are foundational and overarching, providing security and governance across all layers and databases. They are not directly connected to each component but have a dotted line indicating their indirect influence over the entire system, ensuring data integrity and protection against cyber threats.
- Sensor Nodes: LiDAR Sensors, IoT Sensors, and ITS Assets: These nodes are the primary data collection sources. LiDAR sensors capture detailed topographical and infrastructural data. IoT sensors and ITS assets collect various real-time data points, such as traffic flow, weather conditions, and structural health. The data from these sensors feed directly into the central "Data Layer," signifying that the digital twin system relies on continuous, real-time data for accurate representation and analysis.
- External Data Sources: Social Networks and External Open Databases: These elements provide additional dynamic data, such as public sentiment or third-party infrastructural data, to enrich the digital twin's model. The data from these sources enter the same central "Data Layer," emphasizing the system's inclusivity of diverse data types.
- Data Layer: This is the central repository where all incoming data is aggregated. The unification under a "Centralised Data Database" simplifies the system by providing a single point where all data is managed and made ready for processing.
- Data Processing Nodes: AI & ML Algorithms and Big Data Analysis: These processing nodes are crucial for turning raw data into actionable insights. They

apply advanced analytics and machine learning models to find patterns, make predictions, and support decision-making.

- Digital Representation Layer: This layer receives processed data from the AI and ML algorithms. It represents the digital twin model where the virtual representation of the physical infrastructure is maintained and updated. This layer is pivotal for simulation, visualization, and scenario analysis.
- User Interaction Layer: This layer is where the system interacts with end-users. It receives user feedback and provides an interface for users to interact with the digital twin, such as viewing dashboards, generating reports, or controlling simulations.
- Feedback Loops: Two feedback loops are essential for the system's adaptability and responsiveness: Feedback to Infrastructure: Insights from the digital twin model inform updates and adaptations in the physical infrastructure. Feedback to Digital Twin: User interactions and feedback lead to improvements in the digital twin model.

The flowchart is a representation of the continuous and dynamic information flow within a digital twin system. It starts with data acquisition from multiple sources, which then gets centralized and processed. The insights gained are used to update the digital representation, which interacts with users to inform and adapt the physical infrastructure. The inclusion of each component is justified by the need to capture real-time data, secure the system, manage the information effectively, and provide a dynamic model that can evolve based on new data and user interactions. This holistic approach ensures that the digital twin system is a living model, always in sync with its physical counterpart and user requirements.

4 Discussion

The integration of DTs into road infrastructure management presents numerous opportunities and challenges. This section discusses the key findings, lessons learned, and implementation challenges encountered during the development and deployment of the proposed framework.

The implementation of the DT framework significantly enhances decision-making processes by providing real-time data and predictive analytics. This allows for proactive maintenance and efficient traffic management, ultimately leading to improved road safety and reduced congestion. The seamless integration of ITMS and RAMS within the DT framework facilitates comprehensive data sharing and collaboration between different systems. This interconnectedness ensures that all stakeholders have access to the most current and accurate information, enabling coordinated responses to infrastructure issues. Utilizing AI for predictive analytics has proven effective in forecasting maintenance needs. This proactive approach helps in addressing potential road damages before they become critical, thereby extending the lifespan of road assets and reducing maintenance costs.

GIS visualization provides valuable spatial insights, supporting better planning and operational adjustments. The ability to visualize data geographically aids in identifying problem areas and planning maintenance activities more efficiently. The framework's implementation offers substantial socio-economic benefits, including improved traffic flow, reduced travel times, lower maintenance costs, and enhanced road safety. These improvements contribute to the overall quality of life for residents and support economic growth.

The project highlighted several important lessons. Real-time data collection is crucial for the effective functioning of the DT framework. Ensuring that data is continuously updated and accurately reflects current conditions is essential for making informed decisions. The integration of ITMS and RAMS underscores the importance of collaboration between different systems. Effective communication and data sharing between these systems are key to the success of the framework. Additionally, the framework must be scalable and flexible to accommodate future technological advancements and changing infrastructure needs. Building a system that can adapt to new data sources and analytical methods is vital for long-term success. Engaging with all relevant stakeholders, including government officials, road users, and maintenance crews, is also critical. Their input and feedback are invaluable for identifying practical challenges and ensuring the framework meets real-world needs.

Despite its advantages, the framework faces several implementation challenges. Ensuring the privacy and security of collected data is a significant concern. Measures must be implemented to protect sensitive information and prevent unauthorized access. The deployment of advanced technologies such as IoT devices, AI algorithms, and GIS tools requires robust infrastructure and technical expertise. Addressing these technological limitations is essential for the successful implementation of the framework. Integrating various systems and ensuring seamless data flow can be complex. Developing standardized data formats and protocols is necessary to facilitate smooth integration and interoperability. Implementing the DT framework involves significant initial costs and resource allocation. Securing funding and managing resources effectively are crucial for the project's sustainability. Navigating regulatory and policy barriers can also pose challenges. Ensuring compliance with existing regulations and advocating for supportive policies are important for the framework's implementation.

The integration of DTs in road infrastructure management offers substantial benefits, including enhanced decision-making, predictive maintenance, and improved traffic management. However, addressing the implementation challenges is essential for realizing these benefits. By learning from the key findings and lessons learned, future deployments can be more effective, ultimately contributing to more resilient and efficient road infrastructure systems.

5 Methodology

Overview

This section outlines the comprehensive methodology employed to implement Digital Twins in Uzbekistan's road infrastructure, encompassing data collection, analysis, and deployment strategies.

Data Collection

- Sources of Data: Utilisation of both primary sources, such as real-time traffic data from IoT sensors, environmental conditions from weather stations, and secondary sources, including historical traffic data and maintenance records.
- Data Acquisition Techniques: Deployment IoT sensors, satellite imagery, and GPS devices across the road network to collect diverse data sets, including traffic flow, structural conditions of roads, and environmental impacts.
- Data Integration: Aggregation and integration data from disparate sources into a centralized digital twin platform, ensuring data consistency and accessibility.

Analysis Techniques

- Modeling and Simulation: Use advanced simulation software to model road infrastructure and traffic scenarios. This includes using algorithms to simulate traffic flow, environmental impacts, and potential road infrastructure failures.
- Data Analytics: Apply machine learning and AI algorithms to analyze collected data for insights into traffic patterns, predictive maintenance, and infrastructure optimization strategies.

Deployment Process

- Digital Twin Development: Build a virtual model of the road infrastructure that incorporates real-time data feeds, creating a dynamic and accurate representation of the physical world.
- System Integration: Seamlessly integrate the digital twin with existing road infrastructure management systems, allowing for real-time decision-making and enhanced operational efficiency.
- Testing and Validation: Conduct extensive testing, including pilot projects in selected road segments, to validate the digital twin's accuracy and effectiveness.
- Implementation Challenges: Address challenges such as data privacy, technological limitations, and stakeholder resistance through transparent communication, adherence to ethical standards, and showcasing the benefits of digital twin technology.
- Training and Capacity Building: Provide comprehensive training programs for stakeholders, including government officials, infrastructure managers, and technical staff, to ensure they are equipped to utilize the digital twin effectively.

Ethical Considerations

- Address data privacy and security from the outset, ensuring that all data collection and analysis comply with local and international regulations. Engage with stakeholders to foster trust and transparency around the use of digital twin technology.

From a theoretical perspective, the implementation of Digital Twins in Uzbekistan's road infrastructure demonstrates a significant improvement in management, maintenance, and planning processes. By integrating real-time data and predictive analytics, the Digital Twin model facilitates enhanced decision-making, leading to optimized traffic flow, reduced maintenance costs, and improved road safety. The impact of this technology is particularly pronounced given Uzbekistan's diverse geographical and climatic conditions, which pose unique challenges to infrastructure management. Socio-economic benefits include increased efficiency and reduced environmental impact, contributing to sustainable development goals. This case study underscores the potential of Digital Twins to transform road infrastructure management in contexts similar to Uzbekistan's, highlighting the adaptability and scalability of such technologies.

The Fig. 3 illustrates a smart town traffic management system leveraging LoRa and machine learning. Smart sensors gather real-time data, which is transmitted to a LoRa cloud platform for processing. The data undergoes analysis and machine learning to optimize traffic flow, manage smart parking, and enhance public services like emergency response and surveillance. The system aims to reduce traffic congestion, lower pollution, and improve driving experiences by providing adaptive routing and efficient traffic signal management, ultimately enhancing the quality of urban mobility and public safety services.

Integrating a Road Asset Management System (RAMS) with Geographic Information System (GIS) support into an Intelligent Traffic Management System (ITMS) can significantly enhance urban traffic efficiency and road maintenance management.

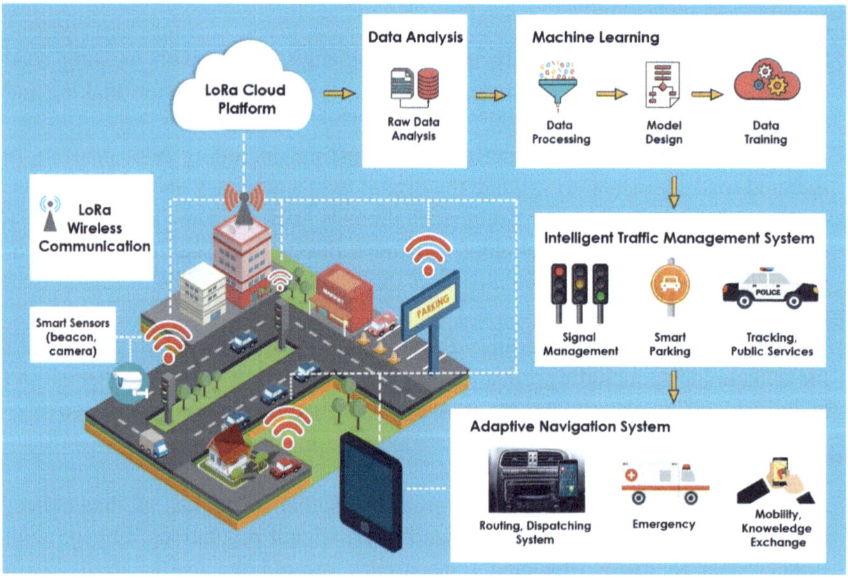

Fig. 3 System overview of LoRa based intelligent traffic management system [14]

The proposed framework involves collecting RAMS data, such as road conditions and maintenance records, and integrating it with GIS to spatially map road assets. This data, collected through smart sensors and LoRa communication, undergoes spatial analysis and visualization to identify traffic patterns and high-maintenance areas. Machine learning algorithms analyze this data for predictive maintenance, optimizing resource allocation and reducing costs. The insights gained enable adaptive traffic management, dynamically adjusting traffic signals, rerouting traffic, and managing smart parking systems to improve flow and safety. A decision support interface, combining GIS visualizations and predictive analytics, facilitates informed decision-making for urban planners and maintenance crews. This integrated approach ensures timely and effective road maintenance, leading to improved urban mobility and safety, and a more resilient transportation infrastructure.

We introduce an advanced digital twin technology designed to revolutionize road asset management and infrastructure maintenance. The technology's development aligns with the development strategy of government and road administration focuses on innovative solutions that enhance urban planning, safety, and sustainability in road networks. Through real-time data analysis, predictive maintenance, and enhanced decision-making capabilities, this technology addresses critical challenges in infrastructure management, offering a scalable, efficient solution for modernizing road systems globally.

The market problem centers around the privatization of road design companies, construction firms, and road material test labs, creating a fragmented industry landscape. This fragmentation results in a scarcity of companies capable of conducting comprehensive network-level road condition surveys. Furthermore, the absence of turnkey software solutions for managing the road sector exacerbates inefficiencies, leading to increased costs, delays, and suboptimal road maintenance and construction quality. This technology aims to bridge these gaps by offering a unified, efficient platform for road sector management, addressing the critical need for integrated, comprehensive road asset management solutions.

The proposed digital twin technology offers a comprehensive solution by integrating real-time monitoring, AI-driven analytics, and GIS for holistic road sector management. It addresses the fragmentation in the industry by providing a centralized platform capable of conducting network-level road condition surveys and analyses. Its innovative aspects include predictive maintenance capabilities, optimization of road design and construction processes, and streamlined data management across privatized sectors.

This turnkey solution facilitates seamless collaboration among road design companies, construction firms, and testing labs, enhancing efficiency and decision-making. Furthermore, the solution incorporates professional training for individuals in the road sector on emerging technologies such as AI/ML and GIS technology. This aims to elevate the professional competence of sector professionals, ensuring they are equipped with the knowledge and skills necessary to leverage the full capabilities of the digital twin technology.

Given Uzbekistan's road administration manages over 43,000 km of roads with an annual budget of approximately 800 million USD for repairs and construction, yet lacks an electronic central database for road conditions and geometry, there's a clear demand for the digital twin technology. The absence of tools for prioritizing road works and inefficient financial distribution underline a significant market need. The commercial potential is substantial, as implementing digital twin technology could optimize the allocation of the annual budget, potentially saving tens to hundreds of millions of dollars through more efficient maintenance scheduling, prioritization of road works, and resource allocation.

The proposed approach envisioned the innovative integration of Digital Twin technology and AI/ML models into Road Asset Management Systems (Fig. 4). This cutting-edge incorporation aims to significantly enhance the system's reliability and the exchange of real-time data. By doing so, we aim to empower road agencies, along with city and local administrations, to transition towards data-driven decision-making processes. This strategic approach is pivotal for adopting proactive measures to address and mitigate traffic disruptions, adapt to climate change, and elevate traffic safety and operations. The integration of Digital Twins and AI/ML models is poised to revolutionize RAMS by enabling the simulation of real-world scenarios with unparalleled accuracy. This capability facilitates predictive analysis and scenario planning, which are critical for the effective maintenance, optimization, and planning of transportation infrastructure. Moreover, the feature of exchanging real-time data ensures that decisions are informed by the latest and most precise data available, thereby optimizing the management of road assets.

The description of each component and their integration within the Digital Twin framework as follows:

Data Collection (📡)

- **ITMS collects data (📡)**: ITMS gathers real-time traffic data using LoRa sensors.
- **Data shared with RAMS (📡)**: This data is shared with RAMS for planning maintenance activities during low-traffic periods.
- **RAMS plans maintenance (📡)**: RAMS uses the data to prioritize maintenance, minimizing traffic disruptions.

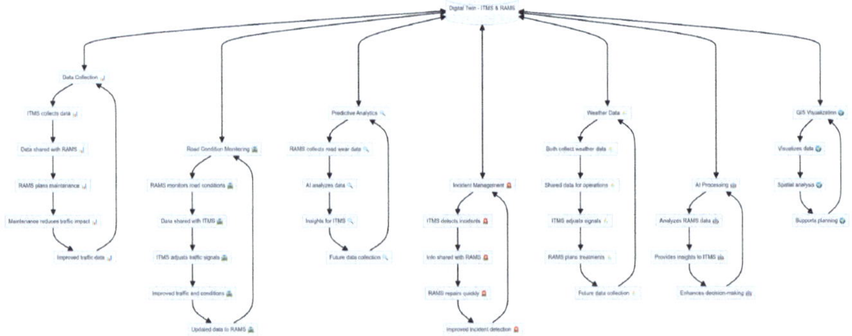

Fig. 4 Integrated digital twin framework for intelligent traffic and road asset management system

- **Maintenance reduces traffic impact (📡)**: Efficient maintenance planning helps reduce traffic impact.
- **Improved traffic data (📡)**: The results feed back to ITMS, enhancing traffic data accuracy.

Road Condition Monitoring (🚧)

- **RAMS monitors road conditions (🚧)**: RAMS uses sensors and IoT devices to monitor road conditions.
- **Data shared with ITMS (🚧)**: This data is shared with ITMS to adjust traffic signals and reroute traffic.
- **ITMS adjusts traffic signals (🚧)**: ITMS dynamically adjusts traffic signals based on road condition data.
- **Improved traffic and conditions (🚧)**: These adjustments improve overall traffic flow and road conditions.
- **Updated data to RAMS (🚧)**: Updated road condition data is fed back to RAMS.

Predictive Analytics (🔍)

- **RAMS collects road wear data (🔍)**: RAMS gathers historical and real-time data on road wear.
- **AI analyzes data (🔍)**: AI algorithms analyze this data, providing insights for ITMS.
- **Insights for ITMS (🔍)**: ITMS uses these insights for proactive maintenance planning.
- **Future data collection (🔍)**: Insights influence future data collection by RAMS.

Incident Management (🔒)

- **ITMS detects incidents (🔒)**: ITMS detects incidents and accidents through surveillance cameras and sensors.
- **Info shared with RAMS (🔒)**: Incident information is fed into RAMS to assess and address road damage.
- **RAMS repairs quickly (🔒)**: RAMS ensures quick repairs and restores normal traffic conditions.
- **Improved incident detection (🔒)**: Restored conditions influence future incident detection by ITMS.

Weather Data (⛈)

- **Both collect weather data (⛈)**: Both ITMS and RAMS collect weather and environmental data.
- **Shared data for operations (⛈)**: This shared data helps both systems adjust operations and maintenance schedules.
- **ITMS adjusts signals (⛈)**: ITMS modifies traffic signal timings during adverse weather conditions.
- **RAMS plans treatments (⛈)**: RAMS plans road treatments based on weather data.
- **Future data collection (⛈)**: Adjusted operations influence future data collection.

AI Processing (□)

- **Analyzes RAMS data (□)**: AI processes data collected by RAMS.
- **Provides insights to ITMS (□)**: AI delivers analytical insights to ITMS.
- **Enhances decision-making (□)**: These insights enhance ITMS's decision-making capabilities for maintenance and traffic management.

GIS Visualization (🗺)

- **Visualizes data (🗺)**: GIS tools visualize data from ITMS and RAMS.
- **Spatial analysis (🗺)**: GIS provides spatial analysis for road conditions and traffic.
- **Supports planning (🗺)**: Spatial insights support planning and operational adjustments for both systems.

The framework ensures a cohesive integration of ITMS and RAMS through shared data, advanced analytics, and visualization, enhancing both traffic management and road maintenance. The integration of ITMS and RAMS within the Digital Twin framework enhances traffic management and road maintenance by leveraging real-time data, AI processing, and GIS visualization. ITMS controls traffic flow, manages parking, and provides public services, while RAMS monitors road conditions and plans maintenance activities. Both systems collect and share data via sensors and IoT devices, which is then analyzed by AI to provide insights and visualized through GIS for effective planning. A pseudo-code for the integrated Digital Twin framework incorporating ITMS and RAMS defines the data flow and processing steps within our system (Fig. 5).

The pseudo-code provides a high-level overview of how the various components interact within the integrated framework, ensuring continuous data flow and real-time adjustments.

The Digital Twin serves as the central node, continually updated with data from ITMS and RAMS, ensuring real-time adjustments and optimal performance. AI processing enables predictive analytics for proactive maintenance, while GIS visualization supports spatial analysis for road and traffic conditions. This integrated approach facilitates seamless data flow, enhancing decision-making and operational efficiency.

This emphasizes the importance of this integration as a major leap towards developing more resilient, safe, and efficient transportation networks. It reflects our commitment to leveraging advanced technologies to improve infrastructure management practices, ultimately contributing to the well-being of the community and the environment. This initiative represents an exciting opportunity to set new standards in the field of road asset management, promising substantial improvements in traffic management, safety measures, and climate resilience. The introduction of a transformative approach to road data collection across Uzbekistan's 43,000 km road network, leveraging low-cost AI/ML algorithms. This innovation represents a significant shift towards more efficient and comprehensive road condition monitoring, providing high-value data essential for optimizing maintenance and repair prioritization. By

```
// Initialize system components
Initialize Digital Twin (DT)
Initialize ITMS
Initialize RAMS
Initialize AI Processing
Initialize GIS Visualization

// Collect and integrate data
While (system is running) {
   // Data Collection
   RealTimeTrafficData = ITMS.collectData(LoRaSensors)
   RAMSData = RAMS.collectData(sensors, IoTDevices)
   WeatherData = collectWeatherData()

   // Share data for integration
   DT.shareData(RealTimeTrafficData, RAMS)
   DT.shareData(RAMSData, ITMS)
   DT.shareData(WeatherData, ITMS, RAMS)

   // Process data with AI
   AIdata = AIProcessing.analyzeData(RAMSData)
   DT.shareData(AIdata, ITMS, RAMS)

   // Visualize data with GIS
   GISVisualization.visualizeData(RealTimeTrafficData, RAMSData,
WeatherData)

   // ITMS Operations
   ITMS.adjustTrafficSignals(RAMSData)
   ITMS.manageTrafficFlow(RealTimeTrafficData)
   ITMS.detectIncidents()

   // RAMS Operations
   RAMS.planMaintenance(RealTimeTrafficData)
   RAMS.monitorRoadConditions()
   RAMS.scheduleRepairs(IncidentData)

   // AI and Predictive Analytics
   MaintenanceInsights = AIProcessing.provideInsights(RAMSData)
   DT.shareData(MaintenanceInsights, ITMS)

   // Update data back to DT
   DT.updateData(ITMS.getUpdatedData())
   DT.updateData(RAMS.getUpdatedData())
}
```

Fig. 5 Pseudo-code for integrating ITMS and RAMS within a digital twin framework

```
// Function Definitions
Function ITMS.collectData(sensors) { ... }
Function RAMS.collectData(sensors, IoTDevices) { ... }
Function collectWeatherData() { ... }
Function AIProcessing.analyzeData(data) { ... }
Function GISVisualization.visualizeData(data1, data2, data3) { ... }
Function ITMS.adjustTrafficSignals(data) { ... }
Function ITMS.manageTrafficFlow(data) { ... }
Function ITMS.detectIncidents() { ... }
Function RAMS.planMaintenance(data) { ... }
Function RAMS.monitorRoadConditions() { ... }
Function RAMS.scheduleRepairs(data) { ... }
Function AIProcessing.provideInsights(data) { ... }
Function DT.shareData(data, recipient) { ... }
Function DT.updateData(data) { ... }
```

Fig. 5 (continued)

automating data collection and analysis, the technology promises substantial cost savings and improved road safety, addressing a crucial market gap.

This marks a pivotal contribution to new production in the road sector, offering enhanced decision-making tools and resource allocation strategies with a broad impact on infrastructure management. The business strategy employs a market-pull approach, directly addressing the critical need for efficient road data management and prioritization within Uzbekistan's vast road network. By implementing AI/ML algorithms for low-cost, comprehensive road condition monitoring, the strategy targets the existing problem of inadequate road maintenance prioritization and financial allocation inefficiencies. Anticipated benefits include significant cost savings, enhanced road safety, and improved maintenance effectiveness, leading to a markedly better user experience for road users and administrators alike, with a scalable model applicable to other regions facing similar challenges. The digital twin technology, tailored for Uzbekistan's road network, promises a revolutionary shift in road asset management through AI/ML-enhanced data collection and analysis. Its strategic market placement addresses a critical gap, with significant potential for cost savings, safety improvements, and operational efficiency. Positioned to redefine the commercial landscape of infrastructure management, its impact extends beyond immediate market success, setting a new standard for global road sector innovation and sustainability. This initiative not only meets current demands but also anticipates future infrastructure challenges, marking a pivotal advancement in the field.

6 Future Directions

The integration of DTs in road infrastructure management is a promising field that offers numerous avenues for future research and practical implementation. This section outlines potential future directions in both theoretical and practical domains.

Theoretical Directions

Theoretically, future research can delve deeper into the development of advanced AI algorithms and machine learning models tailored specifically for road infrastructure management. These models can be designed to improve the accuracy and efficiency of predictive maintenance, traffic management, and incident detection. The exploration of enhanced data integration techniques. Research will focus on developing new methods for seamless data fusion from diverse sources, such as IoT devices, sensors, and satellite imagery, to create a more comprehensive and real-time digital representation of road networks. Furthermore, theoretical advancements in the application of DTs can investigate the impact of climate change on road infrastructure. This includes modeling and simulating the effects of extreme weather events on road conditions and maintenance needs. Studies will also explore the potential of blockchain technology to enhance data security and transparency within the DT framework, ensuring the integrity and reliability of collected data.

Practical Implications

From a practical standpoint, the implementation of DTs in road infrastructure management can be expanded to include pilot projects in diverse geographic and climatic conditions. These pilot projects can provide valuable insights into the adaptability and scalability of the DT framework in different environments. Additionally, practical efforts can focus on developing user-friendly interfaces and visualization tools that make it easier for stakeholders to interact with the DT system and derive actionable insights. There is also a need for comprehensive training programs to equip engineers, urban planners, and policymakers with the skills required to leverage DT technology effectively. Collaboration with industry partners can facilitate the development and deployment of new sensors and IoT devices tailored to the specific needs of road infrastructure monitoring and management.

Integration with Smart City Initiatives

Integrating the DT framework with broader smart city initiatives represents another significant future direction. This involves creating interconnected systems where road infrastructure management is seamlessly integrated with other urban management systems, such as public transportation, energy grids, and emergency services. Such integration can enhance the overall efficiency and resilience of urban infrastructure, leading to smarter and more sustainable cities.

Policy and Regulatory Frameworks

Finally, the development of supportive policy and regulatory frameworks is crucial for the widespread adoption of DT technology in road infrastructure management. Future work can focus on advocating for policies that promote data sharing, interoperability, and the use of advanced technologies in public infrastructure projects. This includes establishing standards for data collection, processing, and storage to ensure consistency and compatibility across different systems and platforms.

The future of DTs in road infrastructure management lies in a synergistic approach that combines theoretical advancements with practical implementations. By exploring new research directions, implementing pilot projects, and developing supportive policies, the potential of DT technology can be fully realized, leading to more resilient, efficient, and sustainable road infrastructure systems.

7 Conclusion

The comprehensive study presented in the "Roads Infrastructure Digital Twin Case Study: Uzbekistan" encapsulates a transformative approach towards enhancing road infrastructure management through the integration of DTs. This novel paradigm, rooted in the foundational methodologies employed by NASA's Apollo program, has evolved to become a cornerstone in modern infrastructure management, particularly within the realms of urban planning and road asset management. The application of DTs in Uzbekistan's road infrastructure, as detailed in this case study, underscores the potential for a revolutionary shift in managing complex road networks, addressing the unique geographical and climatic challenges prevalent in the region.

The study meticulously outlines the technical feasibility, strategic integration, and potential socio-economic impacts of DTs within Uzbekistan's road sector. A critical analysis reveals that DTs offer a multifaceted solution to the pressing challenges faced by the road infrastructure, including environmental extremities, technological integration hurdles, and socio-economic constraints. The digital twin framework, enriched with real-time data, AI-driven analytics, and GIS capabilities, emerges as a dynamic decision-support tool that transcends traditional management approaches, paving the way for predictive maintenance strategies, optimized traffic flow, and enhanced road safety.

Moreover, the research highlights the pivotal role of DTs in fostering sustainable urban development, with significant implications for economic growth and social welfare in Uzbekistan. The integration of DTs is portrayed not merely as a technological upgrade but as a strategic endeavor aligned with the nation's broader development goals, emphasizing efficiency, resilience, and intelligent management of infrastructure.

The case study delineates a clear roadmap for the adoption of DTs in road infrastructure management, detailing the methodology encompassing data collection, analysis, and deployment strategies. The proposed digital twin system, characterized by its comprehensive data integration and processing capabilities, sets a benchmark for smart road infrastructure frameworks, highlighting the importance of cybersecurity, data management, and user interaction in the digital twin ecosystem.

Looking ahead, the study underscores the necessity for continuous innovation, stakeholder engagement, and capacity building to fully leverage the benefits of DTs. The challenges of data privacy, technological limitations, and resistance from stakeholders are acknowledged, with proposed solutions emphasizing transparency, ethical standards, and the demonstrable benefits of digital twin technology.

The paper presents a compelling narrative on the transformative potential of digital twins in revolutionizing road infrastructure management. The findings and insights derived from this study contribute significantly to the body of knowledge on digital twins, offering a template for their implementation in similar geographic and socio-economic contexts. It is recommended that policymakers, urban planners, and infrastructure managers in Uzbekistan and beyond consider the strategic integration of digital twins as a core component of their infrastructure development and management strategies. Investment in digital twin technology, alongside fostering a conducive ecosystem for innovation and collaboration, will be pivotal in achieving sustainable urban development and enhanced quality of life for citizens. The exploration of digital twins in the context of Uzbekistan's road infrastructure marks a significant step towards realizing the vision of smart, resilient, and efficient transportation networks. As this technology continues to evolve, its adoption and integration into road asset management systems represent an exciting frontier for infrastructure management, promising a future where road networks are not only more intelligent but also more responsive to the needs of the communities they serve.

References

1. Shafto M, Conroy M, Doyle R, Glaessgen E, Kemp C, LeMoigne J, Wang L (2010) DRAFT Modeling, simulation, information technology processing roadmap. Technol Area 11
2. Grieves M (2016) Origins of the digital twin concept. https://www.researchgate.net/publication/307509727_Origins_of_the_Digital_Twin_Concept. Accessed on 2 Jan 2024
3. Vatn J (2018) Industry 4.0 and real-time synchronization of operation and maintenance. In: Safety and reliability—safe societies in a changing world—proceedings of the 28th international European safety and reliability conference, ESREL 2018, pp 681–686
4. Errandonea I, Beltrán S, Arrizabalaga S (2020) Digital twin for maintenance: a literature review. Comput Ind 123:103316. ISSN 0166-3615. https://doi.org/10.1016/j.compind.2020.103316
5. Al-Sehrawy R, Kumar B, Watson R (2021) A digital twin uses classification system for urban planning city infrastructure management. ITcon vol 26, Special issue construction 4.0: established and emerging digital technologies within the construction industry (ConVR 2020), 832–862. https://doi.org/10.36680/j.itcon.2021.045
6. Steyn W, Jvd M, Broekman A (2021) Development of a digital twin of a local road network: a case. J Test Eval. https://doi.org/10.1520/JTE20210043
7. Consilvio A, Hernández JS, Chen W, Brilakis I, Bartoccini L, Di Gennaro F, van Welie M (2023) Towards a digital twin-based intelligent decision support for road maintenance. Transp Res Procedia 69:791–798. ISSN 2352-1465. https://doi.org/10.1016/j.trpro.2023.02.237

8. Gao Y, Qian S, Li Z, Wang P, Wang F, He Q (2021) Digital twin and its application in transportation infrastructure. https://doi.org/10.1109/DTPI52967.2021.9540108
9. Marai OE, Taleb T, Song J (2021) Roads infrastructure digital twin: a step toward smarter cities realization. IEEE Netw 35(2):136–143. https://doi.org/10.1109/MNET.011.2000398
10. Chen K, Eskandari Torbaghan M, Chu M, Zhang L, Garcia-Hernández A (2022) Identifying the most suitable machine learning approach for a road digital twin; a systematic literature review. Proceed Inst Civ Eng. https://doi.org/10.1680/jsmic.22.00003
11. Wu J, Wang X, Dang Y, Lv Z (2022) Digital twins and artificial intelligence in transportation infrastructure: classification, application, and future research directions. Comput Electr Eng 101:107983. ISSN 0045-7906. https://doi.org/10.1016/j.compeleceng.2022.107983
12. Ammar A, Maier F, Catchings R, Nassereddine H, Dadi G (2023) Departments of transportation efforts to digitize ancillary transportation asset data: a step toward digital twins. Transp Res Rec. https://doi.org/10.1177/03611981231166000
13. Chen J, Gugssa M, Yee J, Wang J, Goodin C, Das AR (2023) Framework for digital twin creation in off-road environments from LiDAR scans 12529:125290F–125290F. https://doi.org/10.1117/12.2663632
14. Seo SB, Singh D (2018) Smart town traffic management system using LoRa and machine learning mechanism. https://cmte.ieee.org/futuredirections/tech-policy-ethics/2018articles/smart-town-traffic-management-system-using-lora-and-machine-learning-mechanism/

Synergizing Digital Twin Computing with Urban Entrepreneurship and Tactical Urbanism: A Paradigm for Smart City Evolution

Seyedeh Sedigheh Ghorashi and Bahram Hooshyar Yousefi

Abstract In this chapter we will try to present a conceptual approach, reflected in a model, that investigates the possibilities of digital twin computing within urban entrepreneurship and tactical urbanism which eventually serves theoretically the discourse of Digital Twin Computing in the context of Urban Intelligence. This approach besides its technological improvement aspects, combines theoretical concepts and a model based practical application framework to enhance urban intelligence and community participation in smart city-oriented circumstances. The suggested model is organized into three interconnected stages: Comprehensive urban data analytics, Collaborative Urban Intervention Design, and Adaptive Implementation and Scaling. The features of digital twin computing support each step of urban planning, enabling a resilient and flexible approach. The Comprehensive Urban Data Analysis approach, following the prospect framework, encompasses gathering and analyzing diverse urban data. The key advantage of digital twin computing in this particular situation would be its capacity to analyze and visualize real-time data, providing a detailed awareness of urban dynamics. In the sense of Collaborative Urban Intervention Design, based on data-driven insights, this part promotes a co-design process that involves a variety of stakeholders in the design of urban interventions. This stage leverages the synergistic capabilities of digital twin platforms, integrating urban inhabitants, business owners, city planners, and regulating elements in a collective virtual setting. It would allow stakeholders to examine these scenarios as well as outcomes before executing them physically. The last phase of the concept highlights the practical execution of the co-designed urban interventions, with a particular emphasis on adaptability and scalability. This adaptable strategy safeguards that the urban projects maintain their responsiveness to the ever-changing urban needs and challenges. The aforementioned model presents digital

S. S. Ghorashi (✉)
Faculty of Entrepreneurship, University of Tehran, Tehran, Iran
e-mail: sghorashi@ut.ac.ir

B. H. Yousefi
SIDAM Research Group, Stockholm, Sweden

S. Pourroostaei Ardakani and A. Cheshmehzangi (eds.), *Digital Twin Computing for Urban Intelligence*, Urban Sustainability,
https://doi.org/10.1007/978-981-97-8483-7_11

twin computing as a crucial component in the advancement of smart cities, serving to connect the distinction between theoretical urban planning and actual urban growth.

Keywords Tactical urbanism · Urban entrepreneurship · Digital twin computing · Co-designed · Urban interventions

1 Introduction

With the introduction and assimilation of cutting-edge digital technologies, urban development has undergone a paradigm shift at the start of the twenty-first century. The idea of Digital Twin (DT) Computing, a cutting-edge technology that provides real-time digital replicas of physical environments that are dynamic and interactive, is fundamental to this shift [10]. Within the framework of smart cities, DT redefines urban entrepreneurship and tactical urbanism, acting as a catalyst rather than just a technological tool in the context of urban development. The implications of DT Computing in urban environments extend to a wide array of stakeholders. Urban planners and architects find in DT a robust platform for simulating and analyzing urban designs [17]. Entrepreneurs and business leaders view DT as a gateway to innovative urban business models, harnessing real-time data for enhanced decision-making [21]. Local governments and policymakers see the potential of DT in improving urban governance and public service delivery [14].

The body of current research offers significant new perspectives on the use of DT Computing in urban environments. For example, DT's precise, real-time data and simulation capabilities have been crucial in urban planning and management [2]. Through data analytics, DT technology makes it easier to explore new market opportunities and customer insights in the context of urban entrepreneurship [5]. Inspired by Lydon and Garcia [18], and their emphasis on short-term, community-led urban interventions, the concept of tactical urbanism could possibly find in DT a tool for impact assessment and quick prototyping.

Despite these conceptual advancements, several gaps remain in our understanding of the full potential of DT in urban entrepreneurship and tactical urbanism. The technical specifics of implementing DT in varied urban contexts, particularly in less-developed urban environments, are still under-explored [15]. Moreover, there is a lack of comprehensive models that integrate DT with urban entrepreneurship initiatives and tactical urbanism projects, especially in ways that are scalable and replicable across different urban settings. Research Aims and Objectives are:

1. To explore the technical intricacies and applications of Digital Twin Computing in urban entrepreneurship and tactical urbanism.
2. To develop a comprehensive model that integrates DT with urban entrepreneurship initiatives and tactical urbanism projects.
3. To create a scalable and adaptable framework for DT application in diverse urban environments.

4. To promote a more inclusive and participatory approach to urban development using DT.

For a number of reasons, it is essential to comprehend the technical intricacies and applications of digital technology in urban entrepreneurship and tactical urbanism. First of all, it may result in urban planning and development procedures that are more effective, efficient, and adaptable to the changing demands of the city [2]. Second, it can provide data-driven insights to urban entrepreneurs, enabling them to come up with more creative and sustainable business solutions [5]. Finally, based on the model developed by Lydon and Garcia [18], although they have not mentioned it directly, DT can promote a more inclusive and participatory approach to urban development in tactical urbanism by involving a wider range of stakeholders in the urban design process.

To expand our understanding in this domain, a multifaceted approach is required. This eventually would include conducting empirical studies that examine the application of DT in various urban entrepreneurship and tactical urbanism projects. It also involves, as the focus of the current chapter, developing theoretical models that can be tested and refined through simulations and case studies. Cross-disciplinary collaboration between urban planners, technologists, entrepreneurs, and community activists is essential to explore the full spectrum of DT's applications in urban settings [14]. For such applications and the growth of such research, a fundamental theoretical model is required. A guide model, grounded in the body of existing literature, demonstrates the connections and potential synergies between digital twin, urban entrepreneurship, and tactical urbanism for more intelligent urban management. This chapter concentrates on this model's design and presentation as a result.

In order to theoretically support the discourse surrounding digital twin computing in the context of urban intelligence, we will attempt to present a model that explores the potential of digital twin computing within urban entrepreneurship and tactical urbanism. This approach, besides its technological improvement aspects, combines theoretical concepts and a model-based practical application framework to enhance urban intelligence and community participation in smart city-oriented circumstances. The suggested model is organized into three interconnected stages: Comprehensive Urban Data Analytics, Collaborative Urban Intervention Design, and Adaptive Implementation and Scaling, each of which is discussed in this chapter.

2 Literature Review

The literature shows that the transformative potential of DT Computing in reshaping urban landscapes is becoming increasingly recognized. Smart cities are becoming more and more dependent on DT Computing as a means of enhancing urban life through the use of technology. Infrastructure, socioeconomic, and environmental aspects are all included in the integrated approach to managing urban systems that DT provides [9]. The importance of DT in developing real-time responsive urban environments is emphasized by Kitchin [14], who also notes how this improves resident quality of life and the effectiveness of city services.

Through its combination with tactical urbanism and urban entrepreneurship, DT provides a way forward for more inclusive, productive, and sustainable urban environments. To fully realize this potential, though, a number of obstacles must be removed, including operational and technological ones as well as cross-disciplinary cooperation and fair access to DT technologies. Urban development is presented with a comprehensive approach through the integration of DT Computing with Tactical Urbanism and Urban Entrepreneurship. By coordinating economic growth with community needs and environmental sustainability, this synergy promotes creative, inclusive, and sustainable urban solutions [5]. Lehner et al. [15] have conducted recent studies that demonstrate the potential of this integration to convert urban spaces into environments that are dynamic, interactive, and adaptable. While the literature does not explicitly address the integration of DT with tactical urbanism to promote urban entrepreneurship, DT's ability to boost citizen participation and inform data-driven management choices can be a useful tool in both urban entrepreneurship and tactical urbanism.

Even with its potential, there are a number of obstacles to overcome before DT can be used in urban environments; this include the need to integrate diverse data sources, data privacy, and the existence of a strong, scalable IT infrastructure [10]. In addition, DT adoption in urban planning necessitates a change in conventional planning paradigms as well as stakeholder and urban planner capacity building and training [2]. In order to dive into the topic, first we will review the literature separately and then we will summarize and present the most important research.

2.1 Digital Twin Computing in Urbanism Contexts

In the era of industry 4.0, the digital twin is a technology that is revolutionizing which is defined as "… a dynamic virtual representation of a physical object or system across its lifecycle, using real-time data to enable understanding, learning, and reasoning" this is a broad definition. There is evidence in the literature that the field of urban planning and development is changing due to the influence of Digital Twin Computing, which is a major force behind the next wave of urban innovation. Fundamentally, DT is a merging of the digital and physical realms, enabling real-time digital mapping and simulation of all facets of the urban environment [9]. Using IoT, AI, and big data analytics, this technology builds dynamic urban models that are always learning and changing [2]. Tao et al. [23], for instance, show how DT can be applied to monitoring urban infrastructure, forecasting maintenance requirements, and allocating resources as efficiently as possible. Even though the concept of "twin city" [21] goes beyond the sectoral approach of "smart city" [25] virtual replica city is expected to be an important basis for smart city technologies to be applied or used in real cities by integrating the two components (virtual/digital and human) within a multi-purpose infrastructure that aims to produce new knowledge from data and vice-versa. We found many examples of what could be referred to as "digital twin" or "urban digital twin", virtual city services, operating in Singapore and Seoul, respectively.

While many cities have their own unique smart city projects, some technologically advanced cities are attempting to build a virtual city that is precisely modeled after a real city in a digital space [21]. It is anticipated that this virtual replica city will serve as a crucial foundation for the application or utilization of smart city technologies in actual cities. The whole idea is that a smart city has the characteristic of functioning as a platform that includes a wide range of elements related to citizens' lives [26]. A digital twin is a concept in which a virtual world created based on 3D modeling is connected to the real world, and real data is utilized in the virtual space and fed back to the real world [16]. It permits the performance of very complex analyses and decision making in cities.

DT is also considered as a simulation process for a physical product in a virtual space [9]. The pros and cons of various options can be identified through this simulation before taking an actual action on physical products [25]. Yang and Kim (2021) have mentioned in their article a list of urban digital twin applications around the world which provides us with a list of the 40 applications for urban digital twins that have been researched along with brief summaries of each. Many cases where content overlapped imprecisely or where content hierarchies appeared in different applications were discovered during the investigation process (Yang and Kim 2021). Florida et al. [11] believe that, if properly managed, a city that makes more intelligent decisions and involves all stakeholders could become an exporter of solutions—a so-called innovation machine [11].

The information content of city DTs can be used by autonomous systems to operate and update them, including autonomous driving, shipping, and flight operations. Intelligent Transportation Systems (ITS) and traffic centers are new robotic applications that can be planned and simulated with the help of city DTs [24]. Compared to conventional transport systems, simulations fed by data gathered from sensors placed on roads may improve traffic flow regulation, give end-users more information content and safety, and qualitatively increase interaction between road users [24].

Partially examining the city, human factors are essential to the creation and application of digital twin (DT) systems [21]. Although the use of DT as a base for interactive tools for co-design, training, simulation, and decision-making is becoming more popular, there is not much agreement on how to put practical and simple ways of interacting with DTs into practice.

2.2 Urban Entrepreneurship Through the Lens of Digital Twin Technology

Urban entrepreneurship as a relatively new phenomenon that is defined by a conceptual ambiguity is similar to that of the term "entrepreneurship" in the sense of general approaches in their explanations [22]. The field of urban entrepreneurship focuses on turning obstacles found in urban areas into opportunities to improve communities. Addressing problems like unemployment, crime rates resulting from poverty, and raising citizen income and self-sufficiency are all part of this transformative process. The literature also emphasizes the complex relationships between the birth and development of urban entrepreneurship and variables like crime rates, urban security, transit, internet infrastructure, and legislative frameworks [8]. In alignment with existing literature, urban entrepreneurship emerges as a vital solution for cities seeking self-control by mitigating unemployment-related challenges, generating consistent income streams, and fostering citizen independence [7]. Recognizing the factors influencing urban entrepreneurship becomes crucial, not merely as an academic pursuit but as a practical strategy for fostering growth and achieving sustainable urban development [19]. Shahat demonstrate how DT platforms can be leveraged to develop new business models, like online marketplaces that replicate brick-and-mortar stores. The horizons of urban entrepreneurship could be very wide with the help of DT [21]. From the recognition of urban entrepreneurship opportunities by a wide range of citizens to the use of big data and testing entrepreneurial ideas before implementation on a real scale.

2.3 From Tactical Urbanism to Digital Twin Computing

Rapid and often low-cost, innovative, and creative changes form the nature of the "tactical urbanism movement" [3]. Tactical Urbanism is a grass-roots approach and a common term used in community building and rehabilitation, however, it is increasingly becoming a platform to spread ideas and build a stage for change. The term Tactical Urbanism has become an interesting field of research and practice from 2010–2011, when a group of young urbanists created Tactical Urbanism: "Short Term Action, Long Term Change", which brought together several temporary projects in the public spaces of different cities. The authors define it as small-scale and short-term interventions implemented to inspire long-term changes [18].

The combination of long-term planning with fast and modest transformation strategies, as well as an effective formula for putting good ideas into practice, can be a tool to articulate and stimulate the exercise of citizenship and have a positive impact on quality of life. Tactical Urbanism is the movement that applies rapid, small, and/or temporary projects to demonstrate the possibility and potential for large-scale and long-term changes [18]. By giving new directions to places through rapid changes, Tactical Urbanism creates cities that are more environmentally friendly and often motivates people to rethink their habits through different encounters and exchanges these places make possible, as [3] emphasizes:

> Specific projects of tactical urbanism are said to evolve fluidly in relation to broader shifts in political-economic conditions, institutional arrangements, or coalitional dynamics [6].

Tactical urbanism, characterized by short-term, community-based projects aimed at improving urban areas, finds a powerful ally in DT Computing. DT can assist in planning, visualizing, and testing urban interventions before they are implemented. For instance, DT technologies could enable rapid prototyping of urban designs, allowing for community feedback and iterative design improvements.

2.4 Summary and the Most Significant Points of the Literature Review

The historical evolution of digital twin technology from manufacturing to urban applications illustrates a trajectory of increasing complexity and potential impact. This progression reflects an ongoing expansion of digital twin capabilities, driven by technological advancements and an increasing recognition of their value in diverse contexts. The transition from product-focused applications to the management of complex urban systems indicates the vast potential of digital twin technology in contributing to smarter, more efficient, and responsive cities. The most important researches that have been discussed in this research are listed in the Table 1.

3 Methodology

This research adopts an innovative theoretical approach to exploring the integration of Digital Twin (DT) Computing with Urban Entrepreneurship and Tactical Urbanism. Given the constraints related to empirical data collection and the nascent stage of practical applications of DT in this context, the methodology is designed to be theoretical, model-based, and forward-looking.

The application of DT in Urban Entrepreneurship and Tactical Urbanism is an emerging field, with limited practical examples and empirical data available. So, a theoretical model offers a foundation for future empirical research. The mentioned theoretical modeling allows for flexibility and adaptability in exploring a wide range of scenarios and conditions that could possibly serves as a roadmap for future research, guiding empirical studies and practical applications once the field matures. Drawing from a synthesis of existing literature, the model is grounded in theoretical frameworks related to DT, Urban Entrepreneurship, and Tactical Urbanism. Key theories in urban studies, technology innovation, and business models inform the model's conceptual foundation. The model is architectured to represent the complex interactions between DT technology, urban entrepreneurial strategies, and

Table.1 Summary and the most significant points of the literature review

Topic	Key authors and year	Contributions
Early Conceptualization and Manufacturing Applications	Grieves, M. (2014); Boschert, S., & Rosen, R. (2016)	Introduced the digital twin in product lifecycle management; Discussed simulation, prediction, and optimization of product performance in manufacturing
Transition to Urban Applications	Batty, M. (2013); Kitchin, R. (2014)	Explored the concept of smart cities as urban digital twins; Highlighted the role of digital twins in efficient and citizen-focused urban planning
Technological Advancements Fueling Adoption	Tao, F., et al. (2018)	Examined the integration of IoT, AI, and big data in digital twins; Emphasized the transformation of urban planning and management
Conceptualization and Historical Development	Lydon & Garcia (2015)	Introduced Tactical Urbanism concept; Discussed its historical evolution and role in revitalizing public spaces
Principles and Strategies of Tactical Urbanism	Lydon, Mike, et al. (2015); Brenner (2015)	Outlined key principles and strategies for Tactical Urbanism; Analyzed its impact on urban policy and planning
Integration in Smart City Frameworks	Batty, M. (2018); Kitchin, R. (2014)	Discussed digital twins in smart city development; Highlighted their utility in urban governance and analytics
Role in Urban Analytics and Simulation	T. Kim, I. Bong, S. Lee, S. Lee, G. Hwang, S. Kim, J. Nam, J. Ok, D. Kwon, J. Jung, C. Jung, S. Park, H. Lee, H. Choi, Y. Jang, H. Cho, S. Hwang, and H. Jung (2018)	Explored application in urban data analytics and simulation; Discussed impact on urban design and infrastructure
Case Studies of Digital Twin Implementation	Lim, S., & Kim, M. (2018)	Presented case studies from cities like Singapore; Explored transformative potential in various urban contexts
Technical Challenges and Data Integration	T. Kim, I. Bong, S. Lee, S. Lee, G. Hwang, S. Kim, J. Nam, J. Ok, D. Kwon, J. Jung, C. Jung, S. Park, H. Lee, H. Choi, Y. Jang, H. Cho, S. Hwang, and H. Jung. (2018)	Discussed integration challenges and technical barriers; Highlighted data management and real-time processing difficulties

(continued)

Table.1 (continued)

Topic	Key authors and year	Contributions
Socio-economic and Accessibility Issues	Caragliu, A., & Del Bo, C. (2011) Lehner, Daniel & Pfeiffer, Jerome & Tinsel, Erik-Felix & Strljic, Matthias & Sint, Sabine & Vierhauser, Michael & Wortmann, Andreas & Wimmer, Manuel. (2021)	Investigated digital divide and accessibility issues; Discussed impact on urban equity and social inclusion
Advancements in Technology and Applications	Tao, F., et al. (2018)	Predicted advancements in IoT/AI for digital twins; Explored emerging urban applications
Digital Twins and Sustainable Urban Development	Batty, M. (2018); Caragliu, A., & Del Bo, C. (2011)	Discussed role in sustainable development; Examined contributions to smart city sustainability goals
Challenges and Opportunities in Scaling Digital Twins	Kitchin, R. (2014)	Identified challenges in scaling; Highlighted opportunities in city-wide operations and planning

tactical urbanism initiatives. It addresses the technological, socio-economic, and environmental dimensions of Smart Cities (Fig. 1).

To complement the theoretical model, a potential empirical component is initiated, consisting of case studies and simulations which could be the next step of

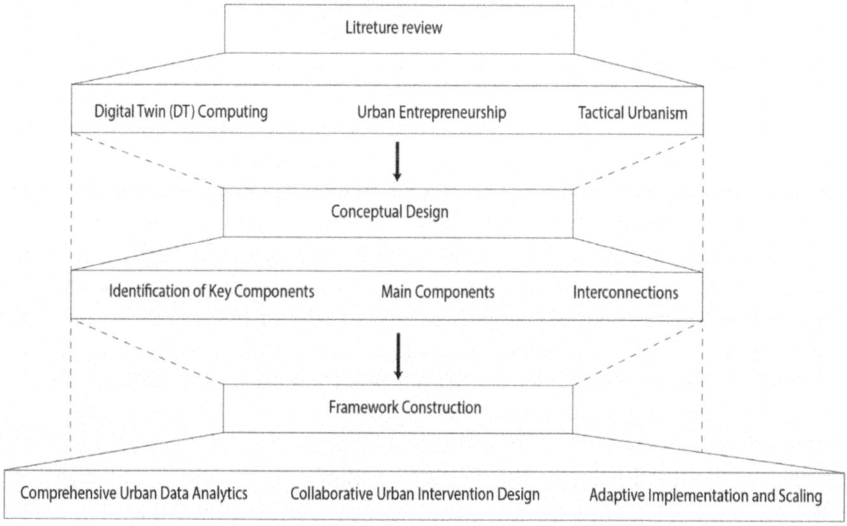

Fig. 1 Conceptual diagram of methodology

this research. The methodology is expected to yield a robust and practical theoretical model that can guide the integration of DT in urban development, particularly focusing on Urban Entrepreneurship and Tactical Urbanism.

4 Discussion and Implications

The central thesis articulated by the authors emphasizes the advocacy for three pivotal dimensions in urban design: entrepreneurially active, technological advanced, and socially inclusive. An astute examination of these domains reveals that the methodologies pertinent to them—urban entrepreneurship, smart city initiatives, and tactical urbanism—are comprehensively addressed within the scope of this research. Moreover, the integration of these dimensions through the application of Digital Twin Technology is posited not merely as feasible, but as a mechanism likely to catalyze synergistic effects among the three areas. This perspective underlines the potential for a transformative impact on urban environments through the confluence of these strategic approaches. Utilizing the conceptual framework provided by Digital Twin Technology, the dimension of technologically advanced urban planning has been reconceptualized as a Comprehensive Urban Data Analysis approach. This transformation underscores a strategic shift towards integrating robust data analytics within urban development paradigms.

The Comprehensive Urban Data Analysis approach commences with an extensive cognification of urban data, facilitated by digital twin technologies. This initial phase involves the collection and analysis of a broad array of urban data, encompassing population dynamics, economic fluctuations, environmental parameters, and spatial utilization. A principal benefit of employing digital twin technology in this context lies in its ability to process and visualize real-time data, thereby furnishing a comprehensive understanding of urban dynamics. This analytical methodology is essential for identifying latent opportunities for entrepreneurial activities and tactical urbanism initiatives, ensuring, eventually, that such projects are grounded in empirical evidence and reflective of contemporary urban conditions. Subsequently, with the help of the idea through the lens of Digital Twin Technology, we have translated socially inclusive into the dimension of Collaborative Urban Intervention Design.

The theoretical framework advocates for a data-driven co-design process that engages a diverse array of stakeholders in the conceptualization of urban interventions. This stage harnesses the synergistic potential of digital twin platforms, enabling a collaborative virtual environment that includes urban residents, business owners, city planners, and regulatory bodies. Within this framework, stakeholders engage in interactive workshops and employ design thinking methodologies to ideate and develop urban intervention scenarios that meld entrepreneurial innovation with the principles of tactical urbanism. The digital twin platform serves as an indispensable toolkit for visualizing, refining, and enhancing these interventions. It permits stakeholders to explore and evaluate these scenarios, along with their potential outcomes, prior to their physical implementation. Digital Twin (DT) technology is pivotal in advancing Tactical Urbanism, offering tools that facilitate community engagement, scenario simulation, and impact assessment. This not only leads to more effective and sustainable urban interventions but also promotes greater community involvement in urban planning processes. By democratizing the planning process, DT technologies ensure that urban interventions are well-aligned with community needs, fostering more inclusive and responsive urban development.

Ultimately, through the application of Digital Twin Technology, we have redefined the concept of "entrepreneurially active" within urban planning as "Adaptive Implementation and Scaling". This transformation involves leveraging digital twins to dynamically adjust and scale urban interventions, facilitating a more responsive and iterative approach to urban development. This adaptive methodology allows for continuous refinement and customization of interventions, ensuring they remain aligned with evolving urban needs and entrepreneurial opportunities.

Digital Twin Computing establishes a robust monitoring and evaluation system that supports continuous real-time assessment and iterative enhancement of urban interventions. This adaptive approach ensures that urban projects remain responsive to fluctuating urban demands and challenges. Upon the successful implementation of these interventions, the framework allows for their expansion or replication across different urban contexts, leveraging the comprehensive data and insights derived from the digital twin platform. Digital Twin technology thus plays a crucial role in fostering entrepreneurship by providing urban entrepreneurs with a sophisticated tool for deeply understanding and interacting with the urban landscape. This encompasses identifying market opportunities, optimizing resource allocation, and innovating urban solutions (Caragliu and Del Bo 2011).

We propose an integrative model for the implementation of digital twins in urban environments, which we have titled the "Comprehensive Urban Analysis and Engagement Framework" (CUAEF). At the heart of CUAEF (Fig. 2) is the utilization of digital twins for thorough urban data analysis, data gathering, and continual data updating. This model facilitates the involvement of various city stakeholders—from urban planners and decision-makers to entrepreneurs—providing them with real-time insights. It also offers the capability to co-design while simultaneously visualizing and receiving feedback on urban projects. This technologically advanced approach is essential for fostering an environment where cities can be entrepreneurially active and socially inclusive.

By leveraging data-driven insights through the "Comprehensive Urban Analysis and Engagement Framework" (CUAEF), our model ensures that urban strategies are aligned with both current and emerging urban trends and conditions. CUAEF underscores the importance of a collaborative urban planning approach, which integrates a diverse array of stakeholders into the planning process. This integration is facilitated by digital twin platforms, which provide interactive tools that allow for the visualization of potential urban changes and scenarios. Such participatory methodologies not only democratize the urban planning process but also enhance civic engagement and stimulate entrepreneurial activities. The fusion of a robust data analysis approach with collaborative urban intervention design fosters the development of innovative urban solutions. These solutions emerge from the identification of market

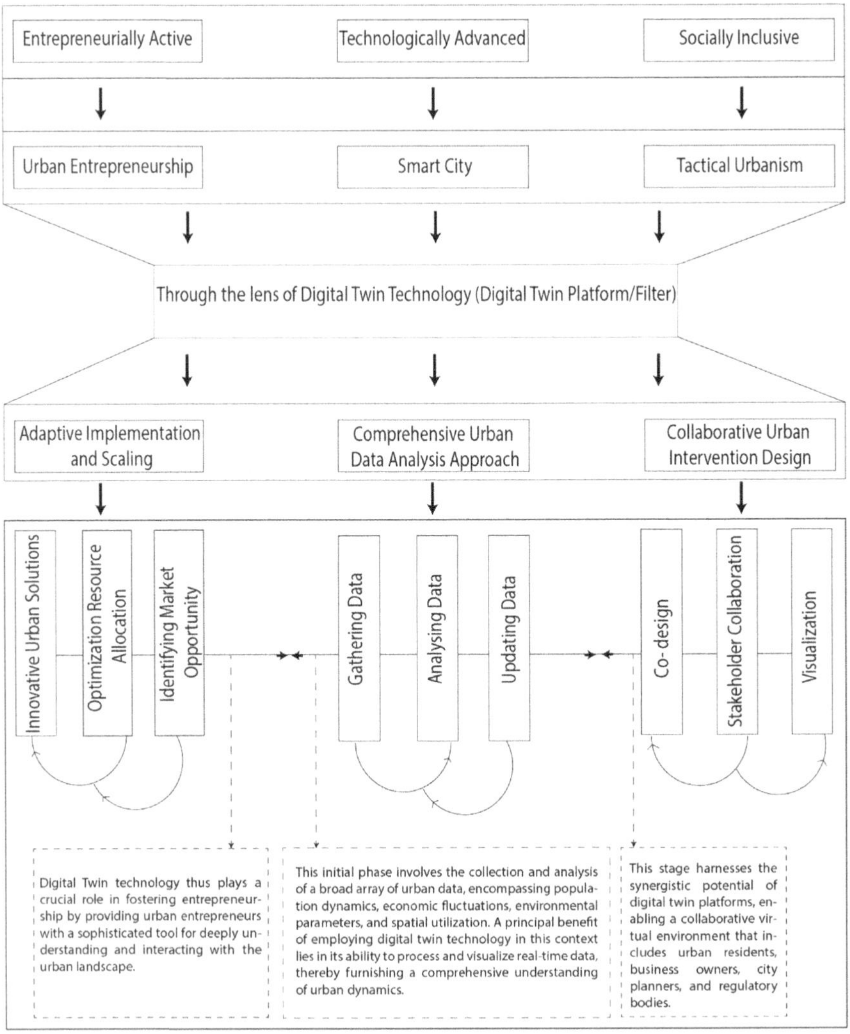

Fig. 2 Comprehensive Urban Analysis and Engagement Framework (CUAEF)

opportunities and the optimization of resource allocations, offering a sustainable and responsive approach to urban development.

The model is designed to integrate seamlessly with advancements in IoT, AI, and other emerging technologies, ensuring its continued relevance and effectiveness over time. Aligned with global sustainability goals, CUAEF underscores the pivotal role of digital twins in promoting sustainable urban development. This includes leveraging digital twins for optimizing resource management, market opportunities identification, and innovative urban solutions.

5 Concluding Remarks

The central thesis articulated by the authors emphasizes the advocacy for three pivotal dimensions in urban design: entrepreneurially active, technological advanced, and socially inclusive. Eventually this chapter, through the lens of Digital Twin Technology, interprets these three critical dimensions—urban entrepreneurship, smart city development, and tactical urbanism—into practical strategies within urban design.

This chapter proposes an integrative model named the Comprehensive Urban Analysis and Engagement Framework (CUAEF), highlighting the critical need to align urban strategies with both current and emergent trends through data-driven insights. The suggested model is organized into three interconnected stages: Comprehensive urban data analytics, Collaborative Urban Intervention Design, and Adaptive Implementation and Scaling. The features of digital twin computing support each step of urban planning, enabling a resilient and flexible approach.

The adaptation of the Comprehensive Urban Data Analysis approach, underpinned by DT technology, establishes a robust framework for technologically advanced urban planning. By leveraging the capabilities of real-time data analysis and visualization, this approach not only illuminates obscured opportunities for entrepreneurial initiatives and tactical urbanism projects but also ensures that these endeavors contain the possibility to be evaluated through solid empirical evidence and to be attuned to the changing urban dynamics. This approach not only supports the strategic deployment of urban interventions but also enhances their effectiveness, adaptability, and relevance to the evolving needs of urban environments.

Within the dimension of Collaborative Urban Intervention Design, DT platforms (as in the model Fig. 2) play a crucial role in facilitating a co-design process that includes a diverse array of stakeholders. By integrating urban residents, businesses, city planners, and regulatory bodies within virtual environments, enabling a collaborative approach to design thinking and scenario visualization. These fosters enhanced community participation in urban planning processes, ensuring that urban interventions are closely aligned with the specific needs and aspirations of the community. In the dimension of Adaptive Implementation and Scaling, DT computing is instrumental in enabling continuous monitoring, evaluation, and iterative improvement of urban interventions. This adaptable strategy is crucial in ensuring that urban projects remain responsive to evolving urban needs and challenges. It provides urban entrepreneurs with sophisticated tools that enhance their understanding of and engagement with the urban environment. Additionally, this approach supports sustainability and scalability, key elements for long-term urban development success.

References

1. Batty M (2013) Big data, smart cities and city planning. Dialogues Hum Geogr 3(3):274–279
2. Batty M (2018) Digital twins. Environment and planning B: urban analytics and city science 45:817–820. https://doi.org/10.1177/2399808318796416

3. Brenner N (2015) Uneven growth: reflections on a curatorial process is "Tactical Urbanism" an alternative to Neoliberal Urbanism?
4. Boschert S, Rosen R (2016) Digital twin—the simulation aspect. In: Mechatronic futures, pp 59–74. Springer, Cham
5. Caragliu A, Del Bo C, Nijkamp P (2011) Smart cities in Europe. J Urban Technol 18:65–82
6. Campo D (2016) Tactical urbanism: short-term action for long-term change. J Urban Des 21(3):388–390. https://doi.org/10.1080/13574809.2016.1167534
7. Cohen B, Muñoz P (2015) Toward a theory of purpose-driven urban entrepreneurship. Organ Environ 28(3):264–285
8. Dana LP, Salamzadeh A, Hadizadeh M, Heydari G, Shamsoddin S (2022) Urban entrepreneurship and sustainable businesses in smart cities: exploring the role of digital technologies. Sustain Technol Entrepreneurship 1(2):100016
9. Deren L, Wenbo Y, Zhenfeng S (2021) Smart city based on digital twins. Comput Urban Sci 1:1–11
10. Deng T, Zhang K, Shen Z (2021) A systematic review of a digital twin city: a new pattern of urban governance toward smart cities. J Manage Sci Eng
11. Florida R, Adler P, Ch, Mellander (2018) The city as innovation machine. https://doi.org/10.4324/9781315143736-8
12. Grieves M (2014) Digital twin: manufacturing excellence through virtual factory replication. White Paper 1–7
13. Kim T, Bong I, Lee S, Hwang G, Kim S, Nam J, Ok J, Kwon D, Jung J, Jung C, Park S, Lee H, Choi H, Jang Y, Cho H, Hwang S, Jung H (2018) A study of smart city strategy in the era of 4th industrial revolution Gyeonggi research institute
14. Kitchin R (2014) The real-time city? Big data and smart urbanism. GeoJournal 79(1):1–14
15. Lehner D, Pfeiffer J, Tinsel E, Strljic M, Sint S, Vierhauser M, Wortmann A, Wimmer M (2021) Digital twin platforms: requirements, capabilities, and future prospects. IEEE Software. https://doi.org/10.1109/MS.2021.3133795
16. Lim S, Kim M (2018) Digital twin application for the success of smart city. KRIHS Policy Brief 687(2018):1–6
17. Liu M, Shuiliang F, Dong H, Xu C (2020) Review of digital twin about concepts, technologies, and industrial applications. J Manuf Syst 58. https://doi.org/10.1016/j.jmsy.2020.06.017
18. Lydon M, Garcia A (2015). A tactical urbanism how-to. https://doi.org/10.5822/978-1-61091-567-0_5
19. Pereyra JAC (2019) Entrepreneurship and the city. Geogr Compass 13(12):e12471
20. Seungho Y, Heechul K (2021) Urban digital twin applications as a virtual platform of smart city. Int J Sustain Build Technol Urban Dev 12(4):363–379. https://doi.org/10.22712/susb.20210030
21. Shahat E, Hyun C, Yeom C (2021) City digital twin potentials: a review and research agenda. Sustainability 13:3386
22. Sriram V, Mersha T, Herron L (2007) Drivers of urban entrepreneurship: an integrative model. Int J Entrep Behav Res 13(4):235–251
23. Tao F, Qi Q, Liu A, Kusiak A (2018) Data-driven smart manufacturing. J Manuf Syst 48:157–169
24. Lehtola V, Koeva M, Elberink S, Raposo P, Virtanen J, Vahdatikhaki F, Borsci S (2022) Digital twin of a city: review of technology serving city needs. Int J Appl Earth Observ Geoinf 114. ISSN 1569-8432. https://doi.org/10.1016/j.jag.2022.102915
25. White G, Zink A, Codecá L, Clarke S (2021) A digital twin smart city for citizen feedback. Cities 110:103064
26. Yang, S, Kim H (2021) Urban digital twin applications as a virtual platform of smart city. Int J Sustain Build Technol Urban Develop 12:636–379. https://doi.org/10.22712/susb.20210030

Navigating the Future with the Nexus Between Digital Twin Computing and Urban Intelligence

Ali Cheshmehzangi and Saeid Pourroostaei Ardakani

Abstract In this concluding chapter of "Navigating the Future", we delve deep into the transformative potential of the nexus between Digital Twin Computing (DTC) and Urban Intelligence (UI). As we have been going through a digital revolution in a rather fast pace, this convergence offers many prospects to revolutionise how we design, manage, and inhabit our urban environments and spaces. Ultimately, the nexus between DTC and UI helps us achieve optimised urban systems and institutions that could nurture urban productivity and sustainability. This chapter serves as a roadmap for the pathways ahead, outlining the key insights and recommendations garnered from contributing explorations and case studies that highlight the importance of integrating design twin computing and urban intelligence. As the name of the book entails, we are eager to see and direct how this integration could shape new paradigms in urbanism, urban management, and urban system development. We begin by reflecting on fundamentals, principles, and case study examples (i.e., both related to technologies and applications), tracing the evolution of both concepts and highlighting the synergies that have emerged between them. From their inception as technological innovations, DTC and UI have evolved into powerful tools for urban planners, policymakers, and citizens alike. Thus, this chapter provides a reflective overview of pathways ahead, by first discussing the importance of this novel nexus and then suggesting a few directions for future research.

Keywords Digital twin computing · Urban intelligence · Digital revolution · Digital age · Digital technologies · Technological innovation · Urban sustainability · Smartness

A. Cheshmehzangi (✉)
School of Architecture, Design and Planning, The University of Queensland, Brisbane, QLD, Australia
e-mail: a.chesh@uq.edu.au

S. P. Ardakani
School of Computer Science, College of Health and Science, University of Lincoln, Lincoln, UK

S. Pourroostaei Ardakani and A. Cheshmehzangi (eds.), *Digital Twin Computing for Urban Intelligence*, Urban Sustainability,
https://doi.org/10.1007/978-981-97-8483-7_12

1 Reflective Overview: Exploring the Intersection of Digital Twin Computing and Urban Intelligence

In this book, most contributions explore case study examples, indicating the diversity of technologies and applications across the board. In recent years, numerous case studies have demonstrated the transformative potential of Digital Twin Computing (DTC) in enhancing Urban Intelligence (UI) across various domains. For instance, cities like Singapore and Barcelona have leveraged DTC to create virtual replicas of their urban infrastructure, enabling real-time monitoring and optimisation of energy consumption, waste management, and transportation networks [28, 40, 41]—i.e., enhancing opportunities for smart cities [35]. These case studies highlight the value of DTC in informing data-driven decision-making and improving the efficiency, resilience, and sustainability of urban environments.

As part of the fundamentals, covered in part one of the book, we see excellent opportunities for intersection between DTC and UI. At its core, Digital Twin Computing involves the creation of virtual representations of physical assets, processes, or systems, which are synchronised with real-time data from sensors [30], IoT devices, and other sources [27]. This synchronisation enables a bidirectional flow of information between the physical and digital realms, allowing stakeholders to monitor, simulate, and analyse various aspects of urban life [6, 20, 37, 46]. Fundamentally, DTC relies on advanced technologies such as IoT, cloud computing, and AI to collect, store, and process vast amounts of data, facilitating predictive modelling [39], scenario analysis, and decision support in urban contexts.

As part of overarching principles, we delve into various aspects that need to be considered in the process of design development, planning process, and decision-making. The principles underlying DTC for urban intelligence revolve around interoperability, scalability, and transparency [1, 3, 52]. Interoperability ensures seamless integration of diverse data sources and systems, enabling stakeholders to aggregate, analyse, and visualise data from multiple domains [22]. Scalability is essential for accommodating the dynamic nature of urban environments and scaling up DTC solutions to encompass entire cities or regions [53]. Transparency is critical for building trust among stakeholders [21] and ensuring the responsible use of data and algorithms in decision-making processes. All three are equally important in optimising our urban systems [8, 15, 16, 42, 43, 47], particularly smart urban systems.

As for technologies, the earlier chapters pinpoint and highlight various aspects for designing our urban intelligence. As highlighted extensively, a variety of technologies underpin DTC for urban intelligence, including IoT sensors, cloud computing platforms, edge computing devices, AI algorithms, and simulation tools [31]. IoT sensors provide real-time data on various urban phenomena [14], such as traffic flow [56], air quality, and energy consumption, while cloud computing platforms enable the storage, processing, and analysis of large datasets [57]. For instance, edge computing devices bring computational capabilities closer to the data source, reducing latency and enabling real-time decision-making [2, 26, 50]. AI algorithms, including machine learning and deep learning techniques, enable pattern recognition, anomaly detection, and predictive modelling in urban environments [12]. Simulation tools allow stakeholders to create virtual models of urban systems and simulate different scenarios [17] to inform decision-making and policy formulation.

Lastly, the applications of DTC for urban intelligence are vast and diverse, spanning multiple domains, including transportation, energy, healthcare, and public safety [5, 32, 49]. For example, DTC can be used to optimise traffic flow, reduce congestion, and improve air quality by simulating different traffic management strategies and analysing their impact on urban mobility [17]. In the energy sector, DTC enables predictive maintenance of infrastructure, optimisation of energy distribution networks, and integration of renewable energy sources into the grid [4]. In healthcare, DTC can support personalised medicine, disease surveillance, and outbreak response by analysing health data from wearable devices, electronic health records, and public health databases [44]. Overall, the applications of DTC for Urban Intelligence [33] are limited only by the imagination and creativity of urban innovators and policymakers.

In summary, the intersection of Digital Twin Computing and Urban Intelligence offers a productive ground for innovation and collaboration in addressing the complex challenges facing modern cities. By embracing the fundamentals, principles, technologies, and applications of DTC, cities can harness the power of data-driven decision-making to create smarter and sustainable solutions for urban environments and cities.

1.1 A Summary of the Book: Contributions and Directions

By delving more deeply into the transformative potential of the nexus between Digital Twin Computing (DTC) and Urban Intelligence (UI), we see crossovers, overlaps, and potential integrated models in cities and urban environments. As known widely, DTC refers to a virtual replica or representation of a physical object, system, or process. Initially developed during the NASA Apollo program in the 1970s, digital twins have evolved significantly and are now applied across various sectors such as manufacturing, healthcare, and smart cities. These virtual replicas simulate and predict performance, perform virtual testing, and facilitate scenario simulations [19]. By integrating sensors, data analytics, and artificial intelligence, digital twins enhance efficiency, productivity, and decision-making. On the other hand, UI primarily focuses on leveraging data and technology to optimise urban systems and institutions [7]. It involves collecting and analysing information from various sources within a city [17], such as sensors, social media, and public records. Urban intelligence enables better urban planning, resource allocation, and sustainable development. Here are some key insights from the book:

1. **Synergies Between DTC and UI: From Overlaps to Integration Opportunities**
   - DTC and UI have evolved from technological innovations into powerful tools for urban planners, policymakers, and citizens.
   - The integration of digital twins with urban intelligence allows for real-time monitoring, predictive maintenance, and operational optimisation.
   - By combining data from physical assets (captured by digital twins) with contextual information (provided by urban intelligence), cities can make informed decisions.
2. **Roadmap for the Future: Evolution towards Productivity and Sustainability**
   - Reflecting on fundamentals and principles, relevant chapters of the book trace the evolution of both DTC and UI.
   - Case study examples highlight successful applications of these concepts, and particularly through the potential nexus and integration opportunities.
   - The novel nexus between DTC and UI is crucial for achieving optimised urban systems and nurturing productivity and sustainability.
3. **Directions for Future Research**
   - The case study examples of the book suggest avenues for further exploration, perhaps from ubiquitous models to context-specific solutions.
   - Researchers can investigate how DTC and UI can shape new paradigms in urbanism, urban management, and urban system development; but more importantly, to explore directions through integrated design thinking models.

In summary, the convergence of DTC and UI holds immense promise for creating smarter, more efficient cities [24]. As we navigate the future, understanding and harnessing this nexus [34] will be essential for sustainable urban development [18]. Hence, the nexus between DTC and UI sits very well in the future of smart cities and sustainable communities. We hope this could be directed to revise and reshape our existing planning and design paradigms that are in urgent need of transformative solutions.

1.2 Transformation and Optimisation Opportunities: Beyond Just a Novelty

Drawing on case studies and real-world examples, including the ones included in this book, we illustrate the myriad applications of DTC and UI in enhancing urban resilience, sustainability, and liveability [40]. From predictive modelling to real-time monitoring, these technologies offer unprecedented capabilities for understanding and optimising urban systems [25, 51]. Whether it is optimising traffic flow, managing energy consumption, or enhancing public safety, the possibilities are endless. However, realising the full potential of DTC and UI requires a concerted effort from all stakeholders [29]. We outline a series of recommendations for policymakers, urban planners, industry leaders, and researchers to harness the transformative power of this nexus effectively. This includes investing in infrastructure, fostering interdisciplinary collaboration, and ensuring equitable access to technology.

Moreover, we emphasise the importance of ethical considerations in the deployment of DTC and UI. As these technologies become increasingly integrated into our daily lives, it is necessary to safeguard privacy, security, and inclusivity [13]. By adopting a human-centric approach, we can ensure that the benefits of DTC and UI are realised by all members of society. In conclusion, this book offers a compelling vision for the future of urban development through the power of integrating DTC with UI. By embracing the nexus between the two, we can create new opportunities and platforms for digital-technologies, data-driven, and intelligent pathways for the future of cities, and not just future cities.

2 The Nexus Between Digital Twin Computing and Urban Intelligence

The nexus between Digital Twin Computing (DTC) and Urban Intelligence (UI) represents a convergence of cutting-edge technologies prepared to revolutionise urban development, management, and governance [9, 36, 38, 55]. DTC involves the creation of virtual replicas, or "twins," of physical assets, systems, or processes, enabling real-time monitoring, analysis, and optimisation. On the other hand, UI refers to the integration of various data sources and analytical tools to enhance decision-making and resource allocation in urban environments. Together, these two paradigms offer unprecedented opportunities for understanding, predicting, and shaping the future of cities.

At the heart of the DTC-UI nexus lies the concept of data-driven urbanism [45]. By harnessing the vast amounts of data generated by urban systems, such as transportation networks, energy grids, and environmental sensors, DTC enables the creation of dynamic digital twins that mirror the complexities of their real-world counterparts. These digital twins serve as powerful simulation platforms, allowing planners and policymakers to test different scenarios, evaluate potential interventions, and predict the impacts of various policy decisions. One of the key benefits of DTC is its ability to facilitate predictive modelling and optimisation. By integrating real-time data streams with advanced analytics and machine learning algorithms, digital twins can anticipate future trends, identify emerging risks, and optimise the performance of urban systems in near real-time [11]. For example, in transportation planning,

DTC can be used to simulate traffic flow, predict congestion patterns, and optimise route planning [54] to minimise travel times [17] and reduce emissions.

Furthermore, DTC enables enhanced situational awareness and decision support capabilities for urban stakeholders [23]. By visualising complex urban systems in a virtual environment, decision-makers can gain insights into system dynamics, identify potential bottlenecks or vulnerabilities, and explore alternative strategies for improving resilience and sustainability [10]. For instance, in emergency management, DTC can be used to simulate disaster scenarios, assess evacuation routes, and coordinate response efforts more effectively. Meanwhile, UI complements DTC by providing the analytical framework and decision-making tools necessary to extract actionable insights from the wealth of data generated by digital twins. Through techniques such as data fusion, machine learning, and spatial analysis, UI enables urban stakeholders to extract patterns, trends, and correlations from disparate data sources, ranging from satellite imagery and social media feeds to sensor networks and administrative records.

By integrating DTC with UI, cities can unlock new possibilities for evidence-based policymaking, resource allocation, and service delivery. For example, in urban planning, digital twins can be used to simulate the impacts of land use policies, zoning regulations, and infrastructure investments on factors such as housing affordability, economic development, and environmental quality [48]. Meanwhile, UI tools can help planners analyse demographic trends, socioeconomic disparities, and community preferences to inform more inclusive and equitable decision-making processes.

Moreover, the DTC-UI nexus has the potential to empower citizens and communities by democratising access to information and participatory decision-making processes. By providing stakeholders with access to digital twins and UI tools, cities can foster greater transparency, accountability, and civic engagement. For instance, citizens could use interactive dashboards and simulation platforms to explore different development scenarios, provide feedback on proposed projects, and collaborate with policymakers to co-create more sustainable and liveable urban environments. In conclusion, the nexus between Digital Twin Computing and Urban Intelligence represents a paradigm shift in how we understand, plan, and manage cities. By combining the predictive capabilities of DTC with the analytical tools of UI, cities can harness the power of data to address complex urban challenges and create more resilient, sustainable, and equitable communities. However, realising the full potential of this nexus will require interdisciplinary collaboration, investment in data infrastructure, and a commitment to ethical and inclusive urban governance.

3 Pathways Ahead and Navigating the Future Through Integrated and Digital Thinking

Integrating digital twin computing with urban intelligence, as evidenced throughout the book, can revolutionise city management and planning. Here are six pathways (Fig. 1) to achieve this integration:

1. **Real-Time Monitoring and Simulation**: Develop digital twins of urban environments that continuously collect data from various sensors placed throughout the city. Integrate this data with urban intelligence systems to provide real-time insights into traffic flow, energy consumption, air quality, and other key urban metrics. This allows for predictive modelling and simulation of various scenarios to optimise city operations.
2. **Predictive Maintenance and Asset Management**: Use digital twins to create virtual replicas of critical infrastructure such as bridges, roads, and utility networks. By integrating these digital twins with AI-driven predictive maintenance algorithms, cities can anticipate and prevent infrastructure failures, minimising downtime and optimising maintenance schedules.
3. **Citizen Engagement and Co-Creation**: Leverage digital twins to create immersive virtual environments that enable citizens to explore and interact with their city in new ways. Integrate these virtual environments with urban intelligence platforms to gather feedback from citizens and involve them in the co-creation of urban solutions, fostering greater civic engagement and participation.
4. **Environmental Sustainability and Energy Efficiency**: Use digital twins to model the environmental impact of urban development projects and integrate them with urban intelligence systems to identify opportunities for energy savings and emissions reductions. By analysing data on energy consumption,

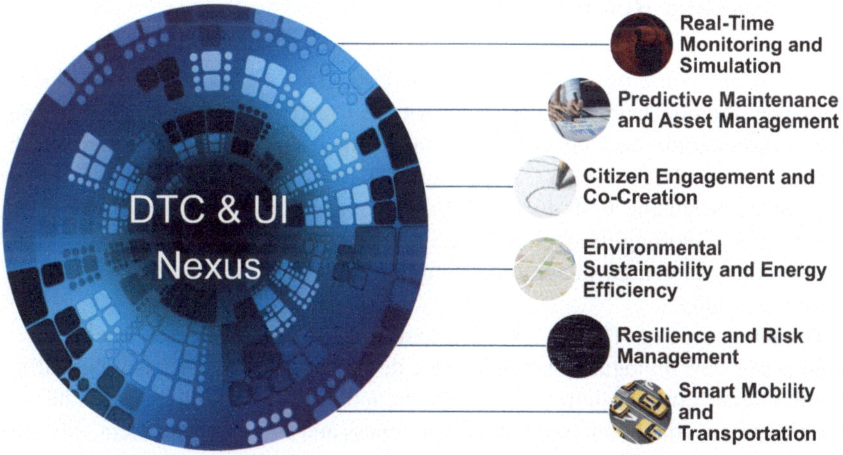

Fig. 1 Six potential pathways for innovation through DTC and UI integration . *Source* The Authors

waste generation, and greenhouse gas emissions, cities can implement targeted interventions to promote sustainability and combat climate change.

5. **Resilience and Risk Management**: Develop digital twins that simulate various disaster scenarios, such as floods, earthquakes, or pandemics, and integrate them with urban intelligence systems to assess vulnerabilities and develop resilience strategies. By analysing real-time data and running predictive simulations, cities can better prepare for and respond to emergencies, minimising damage and saving lives.
6. **Smart Mobility and Transportation**: Create digital twins of transportation networks, including roads, public transit systems, and parking facilities, and integrate them with urban intelligence platforms to optimise traffic flow, reduce congestion, and improve transportation efficiency. This can involve deploying smart sensors, cameras, and AI algorithms to monitor and manage traffic in real-time, as well as providing personalised mobility services to citizens.

Central to all, we urge for dynamic urban planning and design that could help reshape our living and/or built environments. By utilising digital twins, particularly through DTC, we can create dynamic 3D models of urban environments that can be updated in real-time with data from IoT devices, satellite imagery, and other sources. Ultimately, we are able to integrate these models with urban intelligence systems to support data-driven urban planning and design decisions, enabling cities to optimise land use, transportation networks, and infrastructure investments. These pathways represent just a few of the many possibilities for integrating digital twin computing with urban intelligence to create smarter, more sustainable, and resilient cities. By harnessing the power of data, AI, and immersive technologies, cities can unlock new opportunities for innovation and improve the quality of life for their citizens.

4 Concluding Remarks

In the modern era, cities are becoming increasingly complex ecosystems, facing challenges ranging from population growth and resource constraints to environmental sustainability and infrastructure management. In response, innovative technologies such as Digital Twin Computing (DTC) and Urban Intelligence (UI) have emerged as promising tools to navigate the complexities of urban environments. This book has thoroughly explored the synergistic relationship between DTC and UI, examining how their integration can revolutionise urban planning, governance, and sustainability.

The integration of 'Digital Twin Computing' with or into 'Urban Intelligence', could start a new dialogue in how cities are developing, particularly from the urban planning perspective, design of urban systems, and smart city development strategies. As cities are inevitably moving towards smartness and smart development, city officials, governmental units, and decision-makers are working more closely with practitioners who reinvest strategies through digital-based and data-driven approaches.

Such movement would enable the redesign of cities and communities with new outlooks, and hopefully, through new paradigms. By leveraging advanced analytics and simulation techniques, DTC enables predictive modelling, scenario analysis, and decision support, empowering stakeholders to make informed choices and anticipate future challenges. Thus, the move towards enriching UI in practice. As shown throughout the book chapters, we see opportunities that are based on data-driven techniques and technologies. For instance, through AI algorithms, machine learning, and data visualisation techniques, UI enables real-time monitoring, trend analysis, and anomaly detection, enabling cities to respond promptly to emerging issues and optimise resource allocation.

The integration of DTC and UI creates a powerful synergy that amplifies the capabilities of both technologies. By fusing real-time data streams with dynamic simulation models, cities can develop a comprehensive understanding of their complex systems and anticipate the cascading effects of interventions. For instance, urban planners can use DTC-enabled simulations to assess the impact of new infrastructure projects on traffic congestion, air quality, and social equity, while UI tools provide insights into citizen preferences, mobility patterns, and community needs. Furthermore, the combination of DTC and UI facilitates collaborative decision-making and stakeholder engagement, as it enables policymakers, planners, and residents to interact with virtual representations of their city and co-create solutions to pressing challenges. Through participatory modelling, scenario planning, and crowdsourced data collection, cities can harness the collective intelligence of their citizens and build more sustainable communities.

To end the book with a holistic prospect for the future urban sustainability pathways, we hope to highlight the nexus between DTC and UI that holds great promise for navigating the complexities of urbanisation in the twenty-first century. This applies to not only the urbanising contexts or nations but also places where cities are in need of refreshing upgrades, replanning, redesign, and integrated solutions to adopt new urban systems. By embracing a wide range of effective technologies and fostering interdisciplinary collaboration, cities can unlock new opportunities for innovation, sustainability, and human flourishing. However, realising this vision requires addressing technical, ethical, and governance challenges, including data privacy, cyber security, AI generation issues, algorithmic bias, and the growing 'digital divide'. Thus, further research, experimentation, and knowledge-sharing are essential to harness the full potential of DTC-UI integration and create smarter and more equitable cities. Our recent practice imply possibilities for breaking the unsustainable patterns of urban development and shifting towards integrated urban design and planning solutions. With that in mind, we believe digital twin computing could ultimately help us enrich urban intelligence.

References

1. Agrawal K, Aggarwal M, Tanwar S, Sharma G, Bokoro PN, Sharma R (2022) An extensive blockchain based applications survey: tools, frameworks, opportunities, challenges and solutions. IEEE Access 10:116858–116906
2. Al-Dulaimy A, Sharma Y, Khan MG, Taheri J (2020) Introduction to edge computing. Edge Comput Models, Technol Appl Inst Eng Technol London 3–25
3. Al-Farsi S, Rathore MM, Bakiras S (2021) Security of blockchain-based supply chain management systems: challenges and opportunities. Appl Sci 11(12):5585
4. Allahloh AS, Sarfraz M, Ghaleb AM, Al-Shamma'a AA, Hussein Farh HM, Al-Shaalan AM (2023) Revolutionizing Ic Genset operations with iiot and AI: a study on fuel savings and predictive maintenance. Sustainability 15(11):8808
5. Amutha B (2024) Velocious: a resilient IoT architecture for 6G based intelligent transportation system with expeditious movement mechanism. Available from: https://www.researchsquare.com/article/rs-4020623/v1
6. Astarita V, Guido G, Haghshenas SS, Haghshenas SS (2024) Risk reduction in transportation systems: the role of digital twins according to a bibliometric-based literature review. Sustainability 16(8):3212
7. Barns S (2020) Re-engineering the city: platform ecosystems and the capture of urban big data. Front Sustain Cities 2:32
8. Batty M (2024) Digital twins in city planning. Nat Comput Sci 4(3):192–199
9. Calamita K (2024) Towards more meaningful and superior digital experiences in luxury automotive: a roadmap for the introduction of a blockchain technology ecosystem considering the needs of evolving customers. Available from: https://repository.tudelft.nl/islandora/object/uuid:40e897ce-b9d3-40fb-b615-b6768a7fc031
10. Carramiñana D, Bernardos AM, Besada JA, Casar JR (2024) Towards resilient cities: a hybrid simulation framework for risk mitigation through data-driven decision making. Simul Model Pract Theory 133:102924
11. Chen S, Fang J, Chen D, Xu C, Wang Z (2018) Adaptive optimization of sparse matrix-vector multiplication on emerging many-core architectures. In: 2018 IEEE 20th international conference on high performance computing and communications; IEEE 16th international conference on smart city; IEEE 4th international conference on data science and systems (HPCC/SmartCity/DSS). IEEE, pp 649–658
12. Cheshmehzangi A (2021) From transitions to transformation: a brief review of the potential impacts of COVID-19 on boosting digitization, digitalization, and systems thinking in the built environment. J Build Constr Plan Res 9(1):26–39
13. Cheshmehzangi A (2021) Public rights and maintaining the city's social and public services. In: Health U (ed) Sustainability, and peace in the day the world stopped. Springer, Cham, pp 131–137
14. Cheshmehzangi A (2022) ICT, cities, and reaching positive peace. Springer, Singapore
15. Cheshmehzangi A, Batty M, Allam Z, Jones DS (2024) City information modelling: an insight into a new era for the built environment. In: City information modelling. Springer Nature Singapore, Singapore, pp 1–13
16. Cheshmehzangi A, Batty M, Allam Z, Jones, DS (2024) Conclusions and the future of city information modelling (CIM). In: City information modelling. Springer Nature Singapore, Singapore, pp 249–253
17. Cheshmehzangi A, Pourroostaei Ardakani S (2021) Urban traffic optimization based on modeling analysis of sector-based time variable: The case of simulated Ningbo, China. Front Sustain Cities 3:629940
18. Cheshmehzangi A, You N, Siri J, Birch E (2024) The nexus between urban innovation and sustainable development goals (SDGs) In harnessing urban innovation to unlock the sustainable development goals. Springer Nature Singapore, Singapore, pp 1–15

19. Daffodil (2023) How will digital twins evolve in the future? predictions and emerging trends, software engineering insights. Available from: https://insights.daffodilsw.com/blog/the-future-of-digital-twins
20. Dani AAH, Supangkat SH, Lubis FF, Nugraha IGBB, Kinanda R, Rizkia I (2023) Development of a smart city platform based on digital twin technology for monitoring and supporting decision-making. Sustainability 15(18):14002
21. Goldenfein J (2019) Algorithmic transparency and decision-making accountability: thoughts for buying machine learning algorithms. Jake Goldenfein,'Algorithmic transparency and decision-making accountability: thoughts for buying machine learning algorithms' in office of the Victorian Information Commissioner (ed), closer to the machine: technical, social, and legal aspects of AI
22. González Morales LG, Orrell T (2018) Data interoperability: a practitioner's guide to joining up data in the development sector, 60 pp. Available from: https://repository.oceanbestpractices.org/handle/11329/1971
23. Halttula HP, Seppänen O (2022). Situational awareness in construction projects using takt production. In: Annual conference of the international group for lean construction. International Group for Lean Construction (IGLC), pp 164–174
24. Hananto AL, Tirta A, Herawan SG, Idris M, Soudagar MEM, Djamari DW, Veza I (2024) Digital twin and 3D digital twin: concepts, applications, and challenges in industry 4.0 for digital twin. Computers 13(4):100
25. Huang H, Yao XA, Krisp JM, Jiang B (2021) Analytics of location-based big data for smart cities: opportunities, challenges, and future directions. Comput Environ Urban Syst 90:101712
26. Irajifar L, Chen H, Lak A, Sharifi A, Cheshmehzangi A (2023) The nexus between digitalization and sustainability: a scientometrics analysis. Heliyon 9(5)
27. Javaid M, Haleem A, Suman R (2023) Digital twin applications toward industry 4.0: a review. Cogn Robot 3:71–92
28. Kalogeras G, Anagnostopoulos C, Kalogeras G, Anagnostopoulos C, Alexakos C, Monoz L (2021) Digital twins from smart manufacturing to smart cities: a survey. IEEE, Available from https://d1wqtxts1xzle7.cloudfront.net/78189435/09576739-libre.pdf
29. Kuruppuarachchi PM, Rea S, McGibney A (2023) Trusted and secure composite digital twin architecture for collaborative ecosystems. IET Collab Intell Manuf 5(1):e12070
30. López CEB (2021) Real-time event-based platform for the development of digital twin applications. Int J 116(3):835–845
31. MacCarthy BL, Ivanov D (2022) The digital supply chain—emergence, concepts, definitions, and technologies. In: The digital supply chain. Elsevier, pp 3–24
32. Malik H, Anees T, Faheem M, Chaudhry MU, Ali A, Asghar MN (2023) Blockchain and internet of things in smart cities and drug supply management: Open issues, opportunities, and future directions. Internet of Things 100860
33. Malhotra C, Manchanda V, Bhilwar A, Basu A (2021) Designing inclusive smart cities of the future: the Indian context. Solving Urban Infrastruct Prob Using Smart City Technol 631–659
34. Minerva R, Crespi N, Farahbakhsh R, Awan FM (2023) Artificial intelligence and the digital twin: an essential combination. In: crespi N, Drobot AT, Minerva R (eds) The digital twin. Springer, Cham. https://doi.org/10.1007/978-3-031-21343-4_12
35. Mylonas G, Kalogeras A, Kalogeras G, Anagnostopoulos C, Alexakos C, Muñoz L (2021) Digital twins from smart manufacturing to smart cities: a survey. IEEE Access 9:143222–143249
36. Nath SV, Dunkin A, Chowdhary M, Patel N (2020) Industrial digital transformation: accelerate digital transformation with business optimization, AI, and Industry 4.0. Birmingham: Packt Publishing Ltd.
37. Omrany H, Al-Obaidi KM, Husain A, Ghaffarianhoseini A (2023) Digital twins in the construction industry: a comprehensive review of current implementations, enabling technologies, and future directions. Sustainability 15(14):10908
38. Pirayesh A (2022) Innovative integration of auto payment smart contracts with AI and robotics for inspection and construction through BIM: a virtual case study and survey analyzes. Available from: https://www.politesi.polimi.it/handle/10589/211938

39. Pourroostaei Ardakani S, Cheshmehzangi A (2021) Reinforcement learning-enabled UAV itinerary planning for remote sensing applications in smart farming. Telecom 2(3):255–270
40. Pourroostaei Ardakani S, Cheshmehzangi A (2023) Big data-enabled time series analysis for climate change analysis in brazil: an artificial neural network machine learning model. In: Big data analytics for smart transport and healthcare systems. Springer Nature Singapore, Singapore, pp 141–156
41. Pourroostaei Ardakani S, Cheshmehzangi A (2023) Big data analysis for an optimised classification for flight status: prediction analysis using machine learning classifiers. In: Big data analytics for smart transport and healthcare systems. Springer Nature Singapore, Singapore, pp 29–43
42. Pourroostaei Ardakani S, Cheshmehzangi A (2023) Big data analytics and the future of smart transport and healthcare systems. In: Big data analytics for smart transport and healthcare systems. Springer Nature Singapore, Singapore, pp 175–184
43. Pourroostaei Ardakani S, Cheshmehzangi A (2023) Moving forward with big data analytics and smartness. In: Big data analytics for smart urban systems. Springer Nature Singapore, Singapore, pp 111–126
44. Pourroostaei Ardakani S, Cheshmehzangi A (2023) Healthcare infrastructure development and pandemic prevention: an optimal model for healthcare investment using big data. In: Big data analytics for smart transport and healthcare systems. Springer Nature Singapore, Singapore, pp 103–119
45. Pourroostaei Ardakani S, Cheshmehzangi A (2023) Data-driven multi-target prediction analysis for driving pattern recognition: A machine learning approach to enhance prediction accuracy. In: Big data analytics for smart transport and healthcare systems. Springer Nature Singapore, Singapore, pp 63–80
46. Pourroostaei Ardakani S, Xia T, Cheshmehzangi A, Zhang Z (2022) An urban-level prediction of lockdown measures impact on the prevalence of the COVID-19 pandemic. Genus 78(1):28
47. Raslan A, Cheshmehzangi A, Towey D, Tizani W, Kapogiannis G (2024) Impact factor analysis affecting the operation of asset lifecycle by using asset information modeling and blockchain technology. Constr Innov. https://doi.org/10.1108/CI-10-2022-0272/full/html
48. Riaz K, McAfee M, Gharbia SS (2023) Management of climate resilience: exploring the potential of digital twin technology, 3d city modelling, and early warning systems. Sensors 23(5):2659
49. Steri G, Baldini G (2015) The evolution of intelligent transport system (ITS) applications and technologies for law enforcement and public safety. Wireless Public Safety Networks. Elsevier, New York, pp 195–228
50. Su Z, Bentley BL, McDonnell D, Cheshmehzangi A, Ahmad J, Šegalo S, da Veiga CP, Xiang YT (2023) China's algorithmic regulations: Public-facing communication is needed. Health Policy Technol 12(1):100719
51. Suzuki LR (2017) Smart cities IoT: enablers and technology road map In: Rassia S, Pardalos P (eds) Smart city networks Springer optimization and Its applications, vol 125. Springer, Cham, pp 167–190 https://doi.org/10.1007/978-3-319-61313-0_10
52. Vaghani A, Sood K, Yu S (2022) Security and QoS issues in blockchain enabled next-generation smart logistic networks: a tutorial. Blockchain Res Appl 3(3):100082
53. Xu F, Zhang J, Gao C, Feng J, Li Y (2023) Urban generative intelligence (ugi): a foundational platform for agents in embodied city environment. arXiv preprint arXiv:2312.11813
54. Zafar N, Ul Haq I (2020) Traffic congestion prediction based on estimated time of arrival. PLoS ONE 15(12):e0238200
55. Zhang X, Jiang Y, Wu X, Nan Z Jiang Y, Shi J, Zhang Y, Huang X, Huang GG (2024) AIoT-enabled digital twin system for smart tunnel fire safety management. Dev Built Environ 100381
56. Ziafati Bafarasat A, Cheshmehzangi A, Ankowska A (2023) A set of 99 healthy city indicators for application in urban planning and design. Sustain Dev 31(3):1978–1989
57. Zou T, Cheshmehzangi A (2022) ICT adoption and booming e-commerce usage in the COVID-19 era. Front Psychol 13:916843

SPRINGER NATURE

GPSR Compliance

The European Union's (EU) General Product Safety Regulation (GPSR) is a set of rules that requires consumer products to be safe and our obligations to ensure this.

If you have any concerns about our products, you can contact us on ProductSafety@springernature.com

In case Publisher is established outside the EU, the EU authorized representative is:

Springer Nature Customer Service Center GmbH
Europaplatz 3
69115 Heidelberg, Germany

The manufacturer's authorised representative in the EU is Springer
Nature Customer Service Centre GmbH, Europaplatz 3, 69115 Heidelberg,
Germany. If you have any concerns regarding our products, please
contact ProductSafety@springernature.com

Printed and bound by CPI Group (UK) Ltd, Croydon, CR0 4YY
01/10/2026
02225934-0008